Beyond Control

AMERICA'S
THIRD
COAST

Carl A. Brasseaux and Donald W. Davis, series editors

BEYOND CONTROL

The Mississippi River's New Channel to the Gulf of Mexico

James F. Barnett Jr.

University Press of Mississippi / Jackson

This contribution has been supported with funding provided by the Louisiana Sea Grant College Program (LSG) under NOAA Award # NA14OAR4170099. Additional support is from the Louisiana Sea Grant Foundation. The funding support of LSG and NOAA is gratefully acknowledged, along with the matching support by LSU. Logo created by Louisiana Sea Grant College Program.

www.upress.state.ms.us

The University Press of Mississippi is a member of the Association of American University Presses.

First printing 2017

∞

Library of Congress Cataloging-in-Publication Data

Names: Barnett, James F., 1950– author.
Title: Beyond control : the Mississippi River's new channel to the Gulf of
Mexico / James F. Barnett, Jr.
Description: Jackson : University Press of Mississippi, 2017. | Series:
America's third coast series | Includes bibliographical references and
index.
Identifiers: LCCN 2016038729 (print) | LCCN 2016059810 (ebook) | ISBN
9781496811134 (hardback) | ISBN 9781496852113 (trade paperback) |
ISBN 9781496811141 (epub single) | ISBN 9781496811158 (epub institutional) |
ISBN 9781496811165 (pdf single) | ISBN 9781496811172 (pdf institutional) |
Subjects: LCSH: Flood control—Mississippi River—History. |
Floods—Mississippi River Valley—History. | Mississippi
River—Channelization—History. | Floodplain management—Mississippi River
Valley—History. | BISAC: TECHNOLOGY & ENGINEERING / Civil / Flood
Control. | NATURE / Ecosystems & Habitats / Rivers. | HISTORY / United
States / State & Local / South (AL, AR, FL, GA, KY, LA, MS, NC, SC, TN, VA, WV). |
Classification: LCC TC425.M6 B37 2017 (print) | LCC TC425.M6 (ebook) | DDC
551.48/309763—dc23
LC record available at https://lccn.loc.gov/2016038729

British Library Cataloging-in-Publication Data available

For Sharon and the White River voyageurs:
Jim, GG, Lit, George, and the
rest who loved to be on the water

Contents

Acknowledgments

This project would have been impossible without the cooperation of the US Army Corps of Engineers (USACE) and the Mississippi River Commission (MRC). USACE and MRC personnel in Vicksburg, Mississippi; Alexandria, Virginia; and at the Old River Control Complex assisted my research with professional expertise and good-natured enthusiasm. I am indebted to Clyde Joseph "Joe" Harvey, Russell A. Beauvais, and the staff at the Old River control structures and to Morris Oubre and Raymond Harrison at the Old River lock. Jim Dolan patiently and cordially made research material available to me at the US Army Engineer Research and Development Center (USACE ERDC) Library in Vicksburg. Charles A. Camillo, MRC executive director, gave me a copy of his book *Divine Providence: The 2011 Flood in the Mississippi River & Tributaries Project*, 2012, and helped me whenever I emailed with questions. Michael J. Brodhead, John C. Lonnquest, and Larissa Payne at the USACE Office of History sent me scans of the numerous reports and documents that I needed from their collections.

USACE personnel in retirement were also vital to my understanding of the Old River situation. Martin Reuss, whose book *Designing the Bayous: The Control of Water in the Atchafalaya Basin* (2004) was essential to my work, answered my questions at the start of this project about twelve years ago and later assisted me via email. Perry Gustin spent an afternoon with me at his home in Morganza, Louisiana, recounting his experiences at the low-sill control structure the night the wing wall disappeared and took time to answer my questions during subsequent telephone calls. Taylor Frey and her father, Matt Frey, put me in touch with Mr. Gustin.

John M. Barry and Craig E. Colton served as readers for the University Press of Mississippi. Their thoughtful suggestions improved the final manuscript. Carl A. Brasseaux and Donald Davis also read the manuscript and provided helpful suggestions. Beth Richard, Emily Edwards, and Nancy McLemore at Copiah-Lincoln Community College's Willie Mae Dunn Library provided essential and hard-to-find publications through interlibrary loan. Gary Young, Johnny Burt Dixon, John Lee Norwood, and Marianne Fisher-Giorlando helped me trace the route of the Portage of the Cross through the grounds of the Louisiana State Penitentiary.

Cornelia F. Mutel and Robert Ettema graciously shared documents pertaining to their biography of Hans Albert Einstein. I appreciate help from Colleen E. Lyon, University of Texas Library; Denise Grady, *The New York Times*; Judy Bolton and other staff at Louisiana State University's Middleton Library; Jennifer Butt, Nashville's Belmont Mansion; John H. Pardue, Louisiana Water Resources Research Institute; John Wilkinson, Sidney A. Murray Jr. Hydroelectric Power Generator; Mark S. Davis, senior research fellow and director, Tulane Institute on Water Resources Law and Policy; Stanley Richardson, Plaquemine Lock State Commemorative Area; Patricia A. Reid, American Water Resources Association; and Mike Allard, Archives and Record Services Division, Mississippi Department of Archives and History.

Many people offered assistance and provided a helping hand, including Maureen Corcoran, Rebecca M. Anderson, Richard and Jeanie Ray, Lara R. Brown, Martin F. White, Will Richardson, Robert Lee Hadden, Ian W. Brown, Vincas P. Steponaitis, Stanley Nelson, and Oliver A. Houck.

The Historic New Orleans Collection supplied the illustration by Julian Oliver Davidson from *Harper's Weekly*. The map showing Grand Cutoff Bayou is from the Mississippi Department of Archives and History's digital map collection. I had access to numerous reports and newspaper articles thanks to the online resources of the Concordia Parish Library and The HathiTrust. Once again, I am grateful to Craig Gill, Emily Snyder Bandy, Kristi Ezernack, and the staff at University Press of Mississippi. Thanks, too, to Michael Levine for his copyediting work.

Finally, I deeply appreciate the many hours that my wife, Sharon W. Barnett, spent reading and critiquing the book's chapters as they developed. She provided a layperson's perspective and ceaseless encouragement.

Beyond Control

Introduction

April 14, 1973, might have been just another Saturday night in New Orleans. Noisy Bourbon Street nightclubs hawked drinks and dreams to wide-eyed tourists. Three blocks away on Rampart, thousands filled Municipal Auditorium for a concert billed by the Jazz and Heritage Festival as a "Night of Stars." Forgetting for a time about the monster flood descending the Mississippi River, lucky ticket holders escaped into the music of Stevie Wonder, The Olympia Brass Band, and The Ramsey Lewis Trio.[1]

At the time, few in New Orleans had heard of the place called Old River, a US Army Corps of Engineers flood control installation, some 214 miles upriver. If they had heard of it, they would scarcely have believed the possible consequences of the drama unfolding there that evening. For the first time in three thousand years, the Mississippi River above Baton Rouge was seeking a new channel to the Gulf of Mexico.[2] The Crescent City was in peril.

Up at Old River that night, Corps of Engineers personnel stood watch at two adjacent flood control structures, gated weirs with deceptively unpretentious names: low sill and overbank. The low sill spans a 566-foot-wide intake channel from the Mississippi River that feeds into the Atchafalaya River, just seven miles away to the west. The overbank structure sits alongside the low sill. As its name implies, the overbank weir controls floodwaters that overtop the riverbank on the upstream side of the low sill.[3]

The men working on the low-sill control structure could feel the casemate shudder as more than three million gallons of the

Mississippi's floodwaters rushed through its gates every second. Engineers designed the low sill to allow a portion of the Mississippi River to divert into the Atchafalaya Basin while keeping the river's main flow in its historic channel. Only this time, the magnitude of the flood threatened to overpower the engineers' best effort to bridle nature with concrete and steel. In nature's favor is the perilous difference in elevation between the two adjacent rivers. At normal river levels, the Mississippi's surface runs around sixteen feet higher than the surface of the Atchafalaya River. During floods that difference could increase to over 30 feet. Gravity heaved the big river toward the low ground and a much steeper and shorter path to the Gulf of Mexico. The sea is only about 150 miles from Old River via the Atchafalaya Basin, as opposed to the approximately 300-mile route past New Orleans.[4]

The situation became critical when the low-sill's south wing wall, a curved arm extending out into the inflow channel, collapsed under the strain and, where it had stood, the river was rapidly scouring a tunnel through the sediment beneath the structure. Engineers were already aware of a preexisting second scour hole, this one the size of a football field, on the outflow side of the low sill. If the two holes had joined beneath the structure, the low sill would have toppled and the Mississippi River, obeying the inexorable pull of gravity, would have forsaken New Orleans and Baton Rouge to plunge to the Gulf through the Atchafalaya Basin.[5]

Since the city's founding by the French in 1718, the residents of New Orleans have lived with the danger of inundation; however, the possibility of an upriver channel shift threatens something far more menacing than overtopped levees. If the Mississippi River diverts most or all of its flow into the Atchafalaya Basin, the consequences for New Orleans would be catastrophic. Without the river's constant current to hold back the waters of the Gulf of Mexico, New Orleans would become mired in a vast saltwater marsh stretching as far north as Baton Rouge. To the west in the Atchafalaya Basin, the coming of the Mississippi River would be equally disastrous for towns, farms, highways, railways, oil and gas pipelines, and barge canals. At the southern end of the Atchafalaya River, Morgan City would bear the brunt of the deluge.[6]

Thanks to a slice of good luck that April night, the low sill remained functional, though severely damaged, and Mississippi River residents below Old River awoke to enjoy their Palm Sunday, unaware of the previous night's crisis. By June, the Mississippi's waters were falling back toward normal levels and other issues, like the growing political scandal in Washington coincidentally named Watergate, captured the nation's attention. Still, the Mississippi River's drive to change its course did not evaporate with the passing of the 1973 flood. As the muddy water receded, a sinister shadow of insecurity drifted down from Old River to trouble the City That Care Forgot. The flood of 1973, one of the worst on record, raised old concerns in the Lower Mississippi Valley about the ability to control one of the world's mightiest rivers.[7]

Considering the Mississippi River's geologic past, the impending channel change above New Orleans is nothing new. Channel-jumping is one way that the river has adjusted itself periodically in order to carry its load of water and sediment. What is new from the river's perspective is the attempt by humans to hold the Mississippi in one place now that cities and farms have settled behind the levees that line its banks. This artificial confinement creates a dangerous illusion. Engineering has endeavored to transform the Mississippi River into a flowing lake, an iconic American landmark seemingly as unchanging and timeless as the Rocky Mountains.

From an aerial perspective, however, the Mississippi River's turbulent history becomes apparent. Looking down from an airplane window on a clear day over the Mississippi Valley, sunlight glints off of countless U-shaped lakes and sinuous bayous that lie alongside the curving ribbon of the Mississippi. These cutoff lakes and twisting streams were all part of the living river at different times in the past and they reveal the tracks of a watercourse that has been on the move, writhing laterally across its floodplain like a loose garden hose.

In writing about the Mississippi River, it seems natural to use expressions that imbue the legendary waterway with characteristics of a living organism by saying that the river "wants" to change its course and move into the Atchafalaya Basin. Hydrologists turn this around and say that the Atchafalaya may "capture" the Mississippi. In geological jargon, "avulsion" is the word for what happens when a river breaks away from its established route to forge a new channel for itself.[8] Geological circumstances and human interference make Old River the likely location for a major Mississippi River avulsion.

This book tells the story of this short but volatile reach of the Mississippi, beginning with an extraordinary channel reconfiguration around three thousand years ago. The same forces that now compel the river to change direction tore the Mississippi out of its ancient channel on the western side of the valley. Gouging a new course eastward across the floodplain, the river began establishing its present path down to the Gulf. Chapter 1 explores the geological processes surrounding this event and traces the changing dynamics of the Mississippi's confluence with the Red River. Early-eighteenth-century maps of the Mississippi River depict the Old River area as the colonial French found it, marked by a long, gangling meander bend that lassoed both the mouth of the Red River and the distributary (outflow) that is the Atchafalaya River. Not surprisingly, archaeologists have found that this intersection of rivers attracted American Indian settlements. The French maps also document the presence of Indian villages and a historic portage or shortcut used by Indians and French voyageurs to save hours of paddling.

Colonists who came to live in New Orleans soon found that they had to build earthen levees to keep the Mississippi River away from their homes. This communal digging and piling dirt for a common good was reminiscent of the prehistoric Mississippi Valley Indians' ancient practice of constructing ceremonial earthworks. In the case of New Orleans, the arriving Europeans attempted a new approach to riverside settlement that was quite different from the successful strategy followed by the native people of the region. The Indians were expert mound builders, but they didn't use this skill against the Mississippi River. As we will see, there were good reasons why the valley's first inhabitants avoided locating their homes on the banks of the active river channel.

By the end of the eighteenth century, the demographics of the Lower Mississippi Valley and Atchafalaya Basin had changed. In Chapter 2, I look at the Mississippi River's rapidly developing role in the region's burgeoning wealth following the Louisiana Purchase, Indian removal, and the rise of profitable sugarcane and cotton plantations driven by slave labor. With the coming of the steamboat in 1811, the junction of the Mississippi, Red, and Atchafalaya Rivers became a critical intersection for waterborne commerce. But the rivers were dangerous highways. The newly formed Corps of Engineers put Henry Miller Shreve to work removing the snags (uprooted trees) and rafts (logjams blocking waterways) that made river shipping hazardous.

Shreve's arrival in the Lower Mississippi Valley initiated a series of mechanical alterations to the three rivers junction, setting hydrologic forces in motion that could not be reversed. In 1831, Captain Shreve and his herculean snag boat *Heliopolis* severed Turnbull Bend, creating the fragile connection between the three rivers that came to be called Old River.[9] A few years later, amid heated debate that pitted engineers and riverboat pilots against each other, the Louisiana legislature voted to cut off the big meander bend just south of Old River known as Raccourci Bend. The repercussions from these two shortsighted operations were serious enough to stop engineers from cutting off Mississippi meander bends until the 1930s. In tandem with Shreve's cutoff, the state of Louisiana's removal of the tangled raft of trees that blocked boat traffic on the Upper Atchafalaya River further disrupted a natural equilibrium that had existed for three hundred years. Although imperceptible at first, the Mississippi's diversion into the Atchafalaya River gradually increased over time.

For New Orleans, the dire implications of the manipulations at Old River would not become apparent until the mid-twentieth century. Chapter 3 traces the federal government's more urgent calling to prevent flood damage to Lower Mississippi River towns and farms whose privately constructed levees were in ruins following the Civil War. Congress had long resisted spending public funds to build levees protecting private property, but this attitude changed with new engineering studies and the creation in 1879 of the Mississippi River Commission. In a close partnership that continues to this day, the Commission provides civilian and military guidance for the Corps of Engineers' flood control and navigation projects.

One of the Commission's first priorities was a thorough study of the Old River junction. Some constituencies in the Atchafalaya Basin and Lower Red River drainage demanded that the Corps dam Old River to keep the Mississippi's periodic floods from inundating their farmland. Others, including New Orleans interests, were adamant that Old River remain open for both navigation and flood-control purposes. Oblivious to the debate, natural forces were actively reconfiguring the junction's dynamics. Without the Corps of Engineers' intervention, it seemed the Red River would eventually divert all of its flow down the Atchafalaya and close Old River anyway.

While the Mississippi River Commission struggled with the Old River question, a comprehensive report by two military engineers, Captain Andrew A. Humphreys and Lieutenant Henry L. Abbott, set the Lower Mississippi Valley on a course for disaster. Humphreys and Abbott convinced the Commission and the Corps to adopt the policy of containing the Mississippi River and its floods with levees only, as opposed to suggestions by other engineers to employ an array of preventive elements such as reservoirs and artificial outlets. Chapter 4 describes what happened to the infamous "levees only" policy following 1927 and the largest flood in Mississippi River history.

As the United States moved into the Great Depression, the Corps of Engineers launched a massive retooling of Mississippi River flood control, augmenting still higher levees with floodways, reservoirs, and a controversial new campaign of meander bend cutoffs. The Corps' biggest floodway commandeered Old River and the Atchafalaya Basin. Designed to siphon off half of the Mississippi's floodwaters and relieve New Orleans, the Atchafalaya Floodway, in one of Mother Nature's elegant ironies, soon threatened to capture the *whole* river.

A little over a century after Shreve's cutoff and the removal of the Atchafalaya raft, the Atchafalaya River continued to enlarge its capacity, capturing an increasing percentage of the Mississippi's flow. By 1951, Corps engineers predicted full capture of the Mississippi by the Atchafalaya River within the next twenty years. Worse still, the diversion could reach a point of no return by 1960, after which the course change would be impossible to stop. Unwilling to accept the looming doomsday scenario, Congress authorized the Corps of Engineers to take action. Chapter 5 chronicles the efforts to find a way to halt the Mississippi's channel migration and keep the river flowing to New Orleans.

Controlling the water was only part of the puzzle. Management of the Mississippi's complex sediment load was (and remains) a critical issue. Around 150 million tons of sand, silt, and clay move in the river's current every year.[10] This load of mud can block the channels of the Mississippi or the Atchafalaya if allowed to settle in the wrong place. The team of experts assembled to study the problem included Hans Albert Einstein, son of the famous physicist, who was one of the leading authorities on river sediment transport. By the time that the Old River control structures began operating in 1962, the Mississippi was losing between 25 and 30 percent of its flow down the Atchafalaya River. While the Corps could not reverse the diversion, the low-sill structure was able to hold the split at that ratio.

In Chapter 6, I tell the story of Old River and the 1973 flood, the first real test for the control structures. The threat to the low sill, mentioned above, forced the Corps to open the nearby Morganza Floodway control structure, a safety valve at the head of the eastern half of the Atchafalaya Floodway. Completed in 1953, Morganza's gates had never been opened as of 1973, and the Corps' decision to bring the floodway into operation that year was not an easy one. While the diversion of floodwaters might relieve the low sill and lower the flood downriver at New Orleans, the action redirected the menace toward the people and property in the Atchafalaya Basin.

The Mississippi River came frighteningly close to changing its course at the height of the 1973 flood. The river's relentless scouring power revealed near-fatal weaknesses in the low-sill's design, but good fortune kept the Mississippi in its historic channel for the time being. In the wake of the near disaster, three Louisiana State University (LSU) professors published a detailed study of the projected consequences should the Mississippi change its course and surrender to the pull of the Atchafalaya Basin.

Raphael G. Kazmann (Department of Civil Engineering), David B. Johnson (College of Business Administration), and John R. Harris (Department of Civil Engineering) collaborated to present a chilling scenario of a drastically altered Lower Mississippi Valley titled "If the Old River Control Structure Fails? The Physical and Economic Consequences." Chapter 7 examines this important 1980 report with certain aspects updated to reflect today's situation. In the event of a channel diversion at the height of a major flood like that of 1973, the study's hard questions about saltwater contamination of the New Orleans water supply, widespread loss of electrical power, damage to Atchafalaya Basin railway trestles and highway bridges, and interruption of fuel supplies remain to be addressed.

The near-collapse of the low sill forced the Corps to strengthen its control over the Mississippi at Old River. Chapter 8 tracks the Corps' investigation of the low-sill's damages and the development of the massive auxiliary control structure completed adjacent to the low sill in 1986. The flood that nearly destroyed the low sill also inspired the addition of yet another dam-like structure at the Old River Control Complex—a hydroelectric generating station completed in 1990. The mayor of nearby Vidalia, Louisiana, Sidney A.

Murray Jr., watched the floodwaters churn through the low sill and realized that the same action could turn generator turbines. Yet despite the formidable appearance of this battery of installations, the 1973 flood served notice that the unthinkable, an irreversible channel change, was indeed a possibility.

Chapter 8 also looks at the concept of the "100-year flood," a misleading expression that has become an essential tool for floodplain development and a national flood insurance program. Statistical projections about when big floods will occur give the false impression that the Mississippi River is stable and predictable. The truth lies deep beneath the surface of the water. Since the late 1800s, the Corps' river gages have traced the steadily rising riverbed in the vicinity of Old River, the result of sediment accumulation. The changing river flows through a valley that is also transforming in response to geological stress from plate tectonics, faulting, and subsidence. In combination with global climate change uncertainties, these forces will undoubtedly influence the Mississippi's future channel direction.

Chapter 9 begins with a look at a recent Corps of Engineers report that reveals unexpected weaknesses in the hydrology of the Old River Control Complex. The team of engineers completed the study just as the 2011 flood, one of the largest on record, carried enough sediment into the control structure inflow and outflow channels to warrant dredging the Old River system for the first time since its installation.[11] The following year, people in the Lower Mississippi Valley experienced a very different Mississippi River. A drought more severe in places than those of the 1930s Dust Bowl era reduced the river's flow to a fraction of its usual volume, interrupting shipping and allowing Gulf saltwater to threaten New Orleans.

These extreme weather conditions and their effects on the Mississippi River may be a random occurrence or, on the other hand, a possible preview of a future beset by more frequent floods and droughts brought on by climate change. Was the Old River Control Complex designed for a river and environment that no longer exists? If so, the low-sill and auxiliary structures may lose control. The Mississippi seeks the Atchafalaya Basin, and the feared channel jump might happen somewhere else. To end the chapter and the book, I explore this possibility along with some thought-provoking suggestions by professionals about how we might alter our strategy for living with a river we cannot hope to control forever.

The Mississippi River's impending channel diversion at Old River has received some mention in books and magazine articles about Lower Mississippi Valley flood control and the post-Katrina plight of New Orleans; however, the subject has remained marginalized in the mainstream media. In contrast, the earth sciences community has produced numerous research articles and reports about this rare natural phenomenon and its potential consequences.

Despite the enormousness of the task, the Corps of Engineers' control structures continue to hold the Mississippi's flow into the Atchafalaya River at 30 percent. For its part, the Mississippi River takes no notice of the superficial infrastructure within its valley.

The Language of the River

Much of the story about the Mississippi's channel shift and Old River comes from reports written by engineers, geologists, and other scientists. As a historian, my goal is to present this fascinating history to as wide an audience as possible, which means trying to avoid technical jargon. In the chapters that follow, I still found it necessary to use a few technical terms that best describe specific situations, with definitions provided either in the narrative or in an endnote. For the most part, these words and phrases derive from the Corps of Engineers' standard terminology. Two important technical expressions, defined below, appear frequently in many of the chapters because they are a convenient way to describe and compare two of the Mississippi River's basic attributes: the volume or amount of water flowing in the channel and the stage or height of the river's surface.

Cubic Feet Per Second

The Corps of Engineers describes the Mississippi River's volume or "discharge" by "cubic feet per second," the amount of water moving past a given point on the river in one second's time. To illustrate the cubic-feet-per-second measurement, I offer the eighteen-wheeler analogy. The trailer size or capacity for these trucks is about 4,000 cubic feet. Therefore, a flow of one million cubic feet per second equals 250 eighteen-wheeler trailers passing a given point on the river every second.

Another way to describe a volume of water, of course, is in terms of gallons per second, which might be more recognizable to readers (one cubic foot equals 7.48 gallons). But using the cubic-feet expression has its advantages. Besides being compatible with Corps of Engineers documentation, cubic feet makes for a cleaner narrative. Describing the size of a flood using gallons requires larger numbers than are necessary when talking about the same flood's volume in cubic feet. For example, an estimate for the maximum volume of the 1927 flood—2.4 million cubic feet per second—balloons to an unwieldy 17.9 million when converted to gallons per second.

The Corps computes the Mississippi's volume/discharge by multiplying the area of the channel's cross section by the average velocity of the current at that point. The area of the channel is its depth multiplied by its width. Bridges provide a convenient place to measure the channel's cross section, taking depth soundings at intervals along a line measuring the river's width. According to Charles A. Camillo, executive director of the Mississippi River Commission and the Mississippi River and Tributaries Project, the Corps previously used a current meter, a device lowered into the river to measure the speed of the current. Now the preferred measuring instrument is an acoustic Doppler current profiler, which transmits sound waves through the river's current measuring the speed at different depths.[1]

River Stages/Gage Readings

The Corps of Engineers and all other US organizations measuring the water surface height of rivers and lakes use the same system. The Mississippi's surface height is its "stage" or gage reading in feet. The gage reading for a particular location is not a measurement of the river's depth. Mississippi River gages measure the water surface height/stage relative to sea level.

Gaging stations on the Mississippi are progressively higher above sea level as one moves farther inland from the Gulf of Mexico. To keep the stage measurement number in the two-digit range for convenience, "gage zero" for sites north of Old River represents an arbitrarily chosen height above sea level. The river's fluctuating height above the arbitrary gage zero provides the reading for these localities. For example, "zero" on the Mississippi River gage at Natchez, Mississippi, is 17.3 feet above sea level. "Flood stage" at Natchez, when the river can be expected to overtop its banks and begin flooding low-lying areas, is a gage reading of 48 feet (water surface at 65.3 feet above sea level). Some six hundred miles upriver at Cairo, Illinois, gage zero is 270.5 feet above sea level. Flood stage at Cairo is 40 feet (water surface at 310.5 feet above sea level).[2]

In the 1970s, National Geodetic Vertical Datum (NGVD) replaced Mean Sea Level, providing a more accurate base from which to measure vertical distances. In 1988, North American Vertical Datum (NAVD) replaced NGVD.[3] The Corps of Engineers currently gives information on river stages relative to NGVD.[4]

The Corps' operating procedures for issuing river alerts and for activating its floodways and control structures in times of flooding are keyed to the river's discharge rate and stages according to gage readings. These figures also allow the Corps to make comparisons between flood events and track trends such as increasing flood crest levels due to changes in the riverbed.

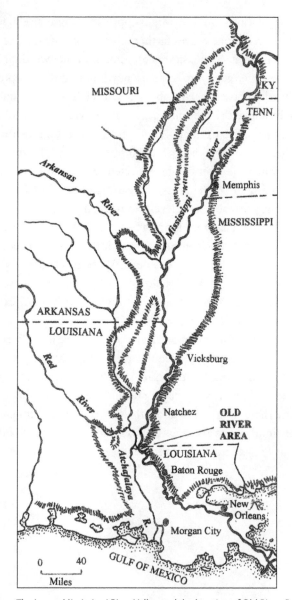

Figure 1.1 The Lower Mississippi River Valley and the location of Old River. Based upon Saucier, *Geomorphology and Quaternary Geologic History of the Lower Mississippi Valley*, Figure 1.

CHAPTER ONE

River of Change

For thousands of years, the Old River area has been a hub of hydrological energy. Here, the Mississippi River's 1.2 million-square-mile drainage basin narrows to a gap less than forty miles wide. The Mississippi, Red, and Atchafalaya Rivers converge and separate within this slender strip of floodplain (see Figure 1.1). Before the Mississippi's channel diversion became a question of "when" instead of "if," the Corps of Engineers' major concern at Old River was facilitating the movement of boats and barges through the junction of waterways. Today, government flood-control installations dominate this out-of-the-way place. The dam-like structures span broad, human-made canals that carry water away from the Mississippi River. The canals seem far too large for the amount of water that usually flows through them.

At the heart of this confluence of rivers, Old River itself is a severed arm of the Mississippi, deliberately cut off from the active channel in the early nineteenth century. (Chapter 2 discusses this engineering project by Henry Miller Shreve.) Now, a quiet slough traces the relict channel where steamboats once churned around a curve known to the river pilots as Turnbull Bend.[1] Cutoffs like this one, both artificial and those that occurred naturally as part of the river's meandering, have created many "old rivers" throughout the Mississippi alluvial valley. These typically horseshoe-shaped, bottomland lakes and sloughs are valuable wetlands, renowned for their fishing and hunting qualities.

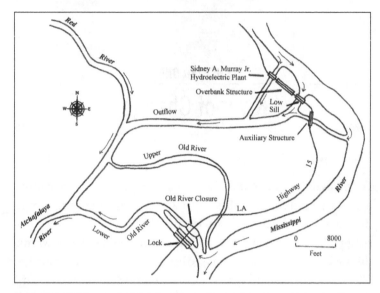

Figure 1.2 Old River control structures and adjacent river channels showing the hydroelectric generator dam (1990), overbank control structure (1959), low-sill control structure (1959), auxiliary control structure (1986), and Old River lock and closure dam (1963). Based upon Reuss, *Designing the Bayous*, 240, 241, 245, 246; Yodis et al., *Geography of Louisiana*, 33, Figure 2.16.

A meeting place of rivers, Old River is also a patchwork of political boundaries. The flood-control complex straddles the 31st parallel, once an international border between the United States and Spanish West Florida from the Mississippi River eastward to the Chattahoochee River. Somewhat confusingly, the edges of four Louisiana parishes lock together here like pieces of a puzzle. The arms of Old River and the connecting Red River serve as the boundaries for the parishes of Concordia, West Feliciana, Avoyelles, and Pointe Coupée. Thanks to Shreve's cutoff at Turnbull Bend, part of West Feliciana Parish, formerly on the east side of the Mississippi, now lies west of the river, noted on maps as Turnbull Island.

Automobile traffic is usually sparse along Highway 15, the two-lane blacktop road that rides the levee down the west side of the Mississippi and bridges the control structures (see Figure 1.2). Like

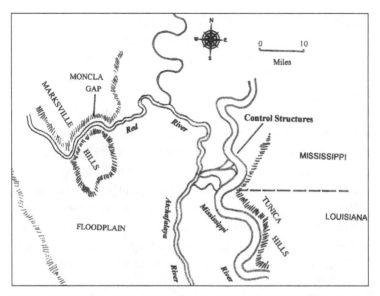

Figure 1.3 The Old River area showing the Mississippi, Red, and Atchafalaya Rivers, Tunica Hills, and Marksville Hills.

the oversize channels behind them, the control structures appear at first glance to be excessive. Towering over these specialized dams, cranes on rolling gantries stand ready to hoist the floodgates and cherry-pick the driftwood that the river piles against the structures. Except for brief periods of high water, the Mississippi River keeps its distance, a hazy half-mile off to the east. Indeed, to a casual traveler passing through, the big river often appears to be too far away and too peaceful looking to warrant the Corps' massive restraining barriers. With the Mississippi's natural connection to the Red and Atchafalaya severed by engineering, the Old River lock, four miles south of the control structures, allows boat traffic to continue moving between the three rivers.

Across the Mississippi from the control structures, the rugged Tunica Hills loom in stark contrast to the floodplain and mark the eastern side of the Mississippi's alluvial valley. Composed of fine-grained loess soil deposited by Ice Age winds, the hills rise over one hundred feet above the bottomland.[2] Angola Prison, Louisiana's state penitentiary, huddles on a levee-protected floodplain at the base of the hills, which form a natural blockade of jungle-covered ridges and deep hollows. On the opposite side of the valley, the Red River flows in from the southern plains. The Red threads through a gap in terraces built from sediments carried south by the ancestral Mississippi in mighty torrents fed by melting glaciers (see Figure 1.3). Variously known as the Marksville Hills and the Avoyelles Prairie, the terraces are around twenty feet higher in elevation than the adjacent alluvial valley. In the narrow floodplain between the Tunica Hills and the Marksville Hills, the Mississippi River's meanderings have continuously reworked more recent sedimentary soils.[3] As we will see, smaller rivers and bayous now occupy the old channel beds left behind by the movements of the Mississippi and Red Rivers through this area.

Old River also marks an important dividing line in the geological history of the Mississippi Valley. To the north, the Mississippi has forged as many as six different courses or meander belts during the last ten thousand years (see Figure 1.4). Below the latitude of Old River during this same period, there have been only two courses down to the Gulf of Mexico. The older of these two meander belts remained active for over four thousand years, stubbornly holding to a path along the western side of the valley. Present-day Bayou Teche flows in this ancient channel, which winds near the cities of Opelousas and Lafayette. This earlier Mississippi River spread out near the coast and emptied into the Gulf across a broad delta stretching from Vermillion Bay eastward to Terrebonne Bay.[4] Figure 1.4 shows the paths of the Mississippi's different meander belts during the past ten thousand years. Younger meander belts cut through and obliterate traces of older courses.[5]

Sometime between five thousand and three thousand years ago, Old River witnessed the rarest of the Mississippi's grand gestures: the abandonment of a long-held meander belt for a new path to the sea. The momentous break with the ancient Teche channel occurred

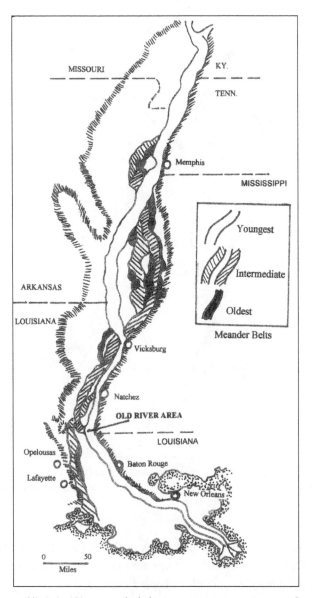

Figure 1.4 Mississippi River meander belts ca. 10,000 years ago to present. Based upon Saucier, *Geomorphology and Quaternary Geologic History of the Lower Mississippi Valley*, Figure 25.

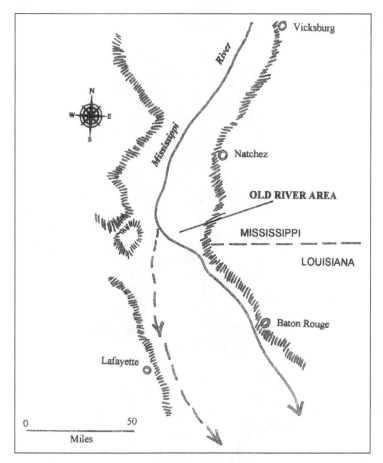

Figure 1.5 Mississippi River diversion into a new meander belt ca. 3,000 years ago. Based upon Saucier, *Geomorphology and Quaternary Geologic History of the Lower Mississippi Valley*, Figure 25, Plates 10, 11.

a few miles northwest of Old River, in the vicinity of present-day Lake Ophelia. A channel shift upriver may have changed the Mississippi's dynamics enough to trigger the fateful crevasse.[6] Whatever the cause, seasonal high water probably overtopped the outside bank of a meander bend, sending floodwater spreading out laterally toward the east. The lower elevation of the floodplain in this direction rapidly captured the entire river. Slicing across the valley through the Old River area, the newborn channel gouged across the bottoms now known as Pomme de Terre Swamp before breaking against the Tunica Hills and turning south.[7] With this prehistoric superflood, the Mississippi began establishing its modern course down the east side of the valley (see Figure 1.5).

Writing about this long-ago channel relocation, geologist Roger Saucier wondered why the change didn't happen sooner: "There is no readily apparent explanation for why more than 4000 years passed before the Mississippi River was able to affect a diversion out of the Teche and its predecessor courses along the western side of the alluvial valley to take advantage of an obvious gradient advantage."[8] The "gradient advantage" to which Saucier referred was not a preexisting characteristic of the landscape. Flowing in the same meander belt for thousands of years, the Mississippi raised its own bed above the surrounding floodplain. The construction agent is the river's load of sediment, particles of sand, silt, and clay carried in the current, flushed into the main channel from numberless upstream tributaries. On average, the river currently carries around 150 million tons of sediment per year; however, it is difficult to say what the sediment load was like in prehistoric times.

The amount of dirt traveling in the Mississippi's current increased dramatically during the nineteenth century, thanks to widespread soil erosion from active land clearing and farming throughout the valley. Record keeping indicates that the river's sediment load peaked at that time and has been decreasing throughout the twentieth century resulting from flood control measures such as reservoirs and levees, which block dirt from reaching the channel. Today, the reduced amount of sediment in the Mississippi may once again approximate the river's natural conditions.[9]

The river's floor rises when the current slows down, allowing heavier sand and silt particles to settle on the bottom. In the Lower Mississippi Valley, two conditions act as brakes to decrease the river's speed and cause riverbed aggradation. One is the elevation of the riverbed relative to sea level. From just above Vicksburg, Mississippi, all the way down to the Gulf, the river's bed lies below sea level, creating a strong drag on the deeper part of the current. The other is the decreasing down-valley slope at the southern end of the Lower Mississippi Valley. The Mississippi's floodplain begins to level off considerably at the latitude of Old River, flattening out and reducing the river's rate of flow over the final three hundred miles to the Gulf.[10]

In tandem with riverbed elevation, periodic overflows increase the height of the Mississippi's natural levees. Floodwaters overtopping the banks deposit the heavier sand and silt particles on the river's shoulders. Before the construction of artificial levees, the flood transported the lighter silt and clay particles away from the channel, creating backswamp deposits. In time, the river's whole meander belt uplifts to form a broad, natural aqueduct. The Mississippi is thus perennially poised above the adjacent bottomland, making channel-jumping crevasses seem inevitable. Yet crevasses that lead to channel diversions are rare.[11]

In seeking to explain why the Mississippi waited more than four thousand years to change its route to the Gulf, geologists found that gradient advantage may not be the most important factor, at least in the river's southern reach. Deep core samples pulled from numerous sites in the Lower Mississippi Valley's alluvial plain by the Corps of Engineers and private energy companies reveal extensive deposits of densely packed soils composed of backswamp silt and clay. When a crevasse occurs in these conditions, the presence of this fine-grained "mud" tends to resist the formation of a new channel.[12]

However, core samples in the Old River region located buried sand deposits associated with abandoned river channels, which can have the opposite effect. In contrast to clay, sand deposits can be highly erodible. If the break in the river's channel occurs near one of these fossil riverbeds, the crevasse can scour out the sand to make room for a new active channel. The presence of an ancestral

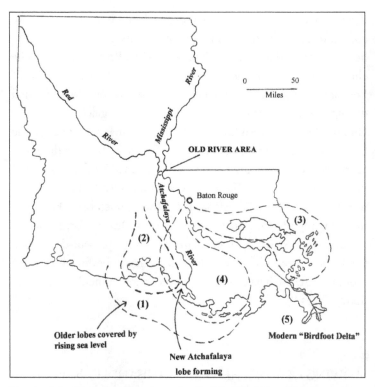

Figure 1.6 Louisiana delta lobes formed by the Mississippi River. Lobes are numbered 1–5 from oldest to youngest. Note new lobe forming at the mouth of the Atchafalaya River. Based upon Yodis et al., *Geography of Louisiana*, 33, Figure 2.19.

Mississippi River channel that ran along the eastern side of the valley over thirty thousand years ago may have facilitated the Teche channel diversion.[13]

In addition to influencing the Mississippi's channel movements, the river's sediment load has shaped Louisiana's distinctive coastline. The finer-grained particles remaining suspended in the river's current are the last to be deposited, sluicing into the marshes near the coast or pluming out the Mississippi's mouth to settle on the sea floor. Over millennia, the sediments discharged from the moving mouth of the Mississippi constructed a series of splayed delta complexes, called lobes, that protrude out into the Gulf. These coastal marshes contrast sharply with the sandy beaches of neighboring Texas and Mississippi. The Mississippi River's ancient Teche channel accounts for the oldest lobes toward the western end of Louisiana's coastal area (see Figure 1.6). These include lobes formed during the rise in sea level at the end of the last glacial epoch. The Gulf of Mexico covered this stretch of Louisiana's ancient coastline before sea level stabilized at its present height about four thousand years ago. To the east are younger lobes formed by the Mississippi in its modern channel. These lobes continue to sink and erode due to subsidence of the Louisiana delta. The characteristic "birdfoot" delta angling out into the Gulf below New Orleans formed about one thousand years ago and marks the river's present-day outlet. Geologists have noted the growth of a new lobe forming at the mouth of the Atchafalaya River, its inception coinciding with the flood of 1973.[14]

◆ ◆ ◆

FLOWING OUT OF THE TEXAS PANHANDLE, the Red River gathers in the Ouachita River drainage system of southern Arkansas and northeastern Louisiana before reaching the Mississippi Valley. Prior to the Mississippi's abandonment of its Teche meander belt, the Red reached the Mississippi through a succession of channels occupied today by bayous Boeuf and Cocodrie, some forty miles southwest of Old River, near present-day Ville Platte. When the Mississippi jumped its channel at Old River and diverted to the east, the Red followed the abandoned Teche riverbed down to the Gulf.[15]

Around 2,800 years ago, the Mississippi completed its shift to the eastern side of the valley to assume the river's modern course from Vicksburg south. The present-day Tensas and Black Rivers follow the channel abandoned by that diversion (see Figure 1.7) The course change left behind a cross-valley swath of erodible sand in

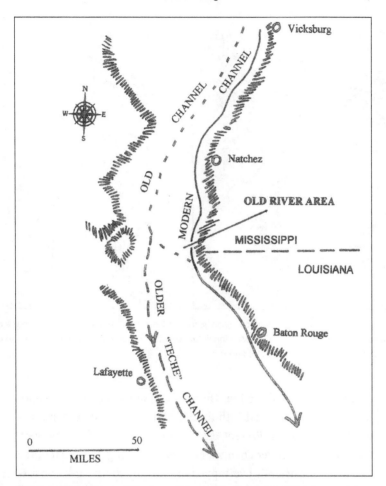

Figure 1.7 The Mississippi River's modern meander belt south of Vicksburg, established ca. 2,800 years ago. Based upon Saucier, *Geomorphology and Quaternary Geologic History of the Lower Mississippi Valley*, Figure 25, Plates 10, 11.

the Old River area where the old riverbed had been. These buried sand deposits marking the old Teche channel diversion played a critical part in the next development at Old River.[16]

A major avulsion pulled the Red River out of its old meander belt and sent it eastward, curving through the broad, sandy path of the

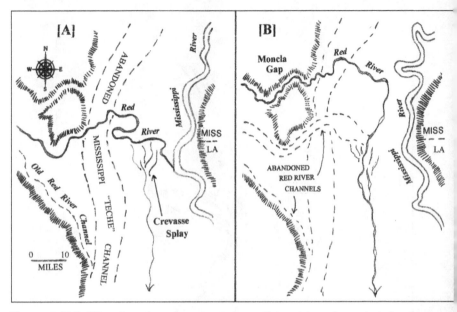

Figure 1.8 (A) Red River channel ca. 2,800–2,000 years ago. Note crevasse splay created when the river overtopped the south bank. (B) Red River channel through Moncla Gap ca. 2,000 years ago. Based upon Aslan et al., "Causes of River Avulsion," Figure 11.

abandoned Teche diversion. The sands served as a conduit, temporarily reuniting the Red with the Mississippi River about ten miles south of Old River. Present-day Bayou des Glaises flows through this former Red River channel. Floodwaters along this reach of the Red River consistently overtopped the stream's south bank, crevassing down into the backswamp swale between the higher ground of the old Teche meander belt and the Mississippi's new eastern channel (see Figure 1.8A). Although this crevassing failed to divert the Red River southward, the sandy sediment and drainage pattern created by these overflows later played an important role in the creation of the Atchafalaya River.[17]

Around two thousand years ago, the Red River avulsed again, this time following an ancient Mississippi channel northeastward through a narrow gorge known as Moncla Gap, just north of present-day Marksville (Figure 1.8B). This new channel scoured across the buried sand of the old Teche meander belt and swung south, flowing down through the swale that paralleled the Mississippi's active channel. The sandy crevasse drainage along the Red's former meander belt helped to direct the Red River down into the Atchafalaya Basin, forming the beginnings of the Atchafalaya River. In this configuration, the Red/Atchafalaya remained separate from the Mississippi, contained within the hollow flanked by the raised meander belts of the Mississippi's active channel and the old Teche channel.[18]

The date of the Red River's avulsion through Moncla Gap comes not from geological sources but from archaeological sites in the vicinity representing prehistoric American Indian occupation. Radiocarbon dating and pottery sequences provide a strong chronology for the prehistory of the Lower Mississippi Valley. Geologists occasionally find this information to be useful in developing time frames for the evolution of river channels. In this case, Indians belonging to the mound-building culture known to archaeologists as Marksville (named for the nearby Louisiana town) built ceremonial mounds at several sites along the course of the Red shortly after its diversion through Moncla Gap. The Marksville culture is securely bracketed in time by the dates 100 BC and AD 400, indicating that the Moncla Gap avulsion occurred before or during this time period.[19]

By the time that these Marksville mound builders settled along the Red River, people had been living in and around the valley of the Mississippi River for more than nine thousand years. The rich environment of the Lower Mississippi Valley attracted hunter/gatherer populations and later agricultural communities; however, the meandering river has erased many archaeological sites in the floodplain. Archaeological investigations in the Old River area reveal evidence of intermittent prehistoric occupation from around 1500 B.C. until historic contact with Europeans.[20]

The archaeology shows us that people have lived close to the Mississippi River for a long time, adjusting to the river's changing meanders and channel diversions. Among the most exceptional archaeological finds are ancient wooden dugout canoes. The dugout is a hollowed-out log typically made from bald cypress and other bottomland trees, fashioned by a tedious process of burning and scraping away of charred wood.[21] These versatile, shallow-draft boats were the key to adaptation in the Lower Mississippi Valley; in a world of water transport, dugouts were the cars and pickup trucks of the pre-industrial age. They transformed a vast network of interconnected rivers, lakes, and bayous into a highway system accessible to anyone with a canoe and paddle.

Long-ago floods sometimes preserved these highly perishable boats through rapid burial in sand and silt, encasing the wood in airtight sediments. Lucky discoveries usually happen when river scouring exposes one of these canoes. Although only a few prehistoric dugout canoes have been found in the region, they undoubtedly represent tens of thousands of such vessels that once plied the Mississippi and its tributaries. With them, generations of native river folk took to the Mississippi's surface as easily as they walked the paths of their villages. Where today the Mississippi River's surface is practically deserted, save for the occasional passing tow and the odd recreational craft, a few hundred years ago the river must have bustled with activity.

Hernando De Soto and his conquistadores found native populations well-adapted to river travel when they traversed the region between present-day Memphis and Vicksburg in the 1540s. The expedition's chronicles describe in vivid detail the populous mound builder chiefdoms and their colorful dugout canoes, some large enough to hold as many as eighty to one hundred warriors. According to expedition accounts, the warrior boatmen were quite expert at negotiating the Mississippi's notoriously challenging cross-channel currents and whirlpools.

De Soto's men had an opportunity to observe the Indians' canoeing skills first hand when the Spaniards released their war horses and took to the Mississippi riding makeshift boats. Their dreams of finding gold and silver dashed, survival depended on a desperate ride down the river to reach the Gulf of Mexico and the safety of Spanish settlements there. The warriors of the river chiefdom called Quigualtam, paddling their giant war canoes, harassed the bearded aliens relentlessly in a running mêlée down the flood-swollen Mississippi in July 1543. Indians in the Old River area likely witnessed the passing of the beleaguered Spaniards and may even have given chase themselves.[22]

◆ ◆ ◆

A MAJOR CHANGE OCCURRED in the hydrological dynamics at Old River around the time of the De Soto expedition. The Mississippi formed a prodigious meander loop that reached out far enough to the west to intercept the Red River, reestablishing a long-lost tributary (see Figure 1.9). The event cannot be dated precisely, but geologists believe that the rejoining of these two rivers occurred sometime in the sixteenth century. The westward meander tracks a gradual adjustment of the Mississippi's bed that may have started as a response to upstream channel fluctuations.[23]

Mississippi meander loops form through a process of point bar extension and bank caving, driven by the sediment load transported in the current. As the river flows around a bend, the current moves faster on the outer flank of the curve, chiseling away at the bank. In tandem with bank erosion on the outside of the bend is the development of a point bar on the inside of the curve, where the current runs more slowly. As discussed above, in a slow current, heavier sand and silt particles settle on the bottom. In the inside curve of a meander bend, these sediments create an ever lengthening sandy bar that mirrors the caving bank on the outside of the bend (see Figure 1.10). The annual rise and fall in the volume of water and sediment transported by the Mississippi causes the point bars to develop incrementally, giving the bars an undulating surface of alternating ridges and swales.[24]

The Mississippi's sixteenth-century reconnection with the Red gave the larger river the opportunity to divert into the Atchafalaya

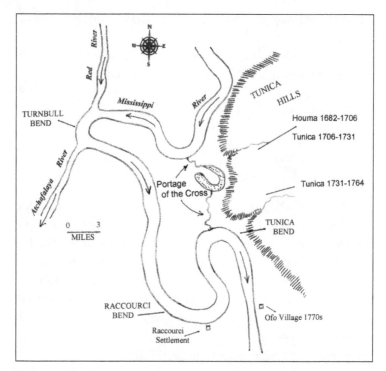

Figure 1.9 The Old River area ca. 1500–1831 showing the mouth of the Red River, the Atchafalaya distributary, approximate locations of Indian villages, the three bends, Raccourci settlement, and the Portage of the Cross. Based upon Aslan et al., "Causes of River Avulsion," Figure 11; Brain, *Tunica Archaeology*, 152; Broutin, "Carte particulière du cours du fleuve Missisipy" (Map 1731); and Gauld (Map 1778), "A Plan of the coast of part of West Florida & Louisiana."

Basin; however, conditions five hundred years ago were not as favorable for avulsion as today's precarious combination of gradient advantage and erodible sand at Old River. The Red River's channel below the new confluence was only large enough to accept a portion of the combined Mississippi and Red Rivers, resulting in the formation of the Atchafalaya distributary off of the main channel about three miles downstream from the mouth of the Red River (see Figure 1.9).[25]

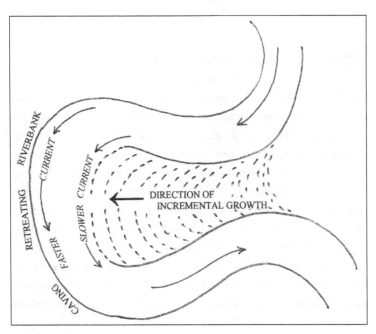

Figure 1.10 Idealized illustration of Mississippi River point bar formation. Based upon Saucier, *Geomorphology and Quaternary Geologic History of the Lower Mississippi Valley*, 109–12, 191–94, Figures 20 and 46.

The Atchafalaya is one of several distributaries or outlets through which the Mississippi has divided its flow over the last several thousand years. Although Mississippi Valley geology indicates that the river has occasionally left its main channel to divert its full flow into a distributary, such wholesale changes are rare and none are documented historically, that is, until the Atchafalaya began in earnest to capture the Mississippi in the twentieth century. Some relatively recent Mississippi River distributaries that formed and later lost their connection to the Mississippi are marked by the present-day St. Francis River in northeast Arkansas, Bayou Macon in southeast Arkansas and northeast Louisiana, Deer Creek in Mississippi's Yazoo Basin, and the Little River in northeast Louisiana.[26]

◆ ◆ ◆

WITH THE RED REESTABLISHED AS A TRIBUTARY and the Atchafalaya River as a distributary, the Mississippi's far-flung meander bend at Old River attained a remarkable state of equilibrium that lasted until the human-made cutoffs in the early nineteenth century. The first accurate maps of the region drawn by colonial French cartographers record the twists and turns of this enormous kink in the Mississippi's course (see Figure 1.9). In time, riverboat pilots assigned names to the three horseshoe-shaped curves that make up the meander loop. The northernmost curve intersecting the Red and Atchafalaya Rivers became Turnbull Bend, named for a prominent landowner with riverside property. South of Turnbull Bend was Raccourci Bend, identified with a small community at the meander's southernmost reach. The meander resumed its southward direction after looping around Tunica Bend, which took its name from the adjacent hills and the Indian tribe that once lived nearby. With all three bends combined, the meander was thirty-seven miles long.

At Tunica Bend, travelers going upriver might have been disconcerted to find themselves paddling due south for the next several miles before rounding Raccourci Bend and heading north again. The French maps showing this reach of the river also reveal an important eighteenth-century colonial landmark, the Portage de la Croix (Portage of the Cross). The famous portage followed a swampy water system that cut across the neck of the meander, a distance of only about six miles, saving travelers from having to paddle all the way around the tortuous circuit.

Voyageurs approaching from the south entered the portage via the mouth of a bayou marked by the French with a large wooden cross planted on the riverbank. Without the cross to show the way, this stream would have been difficult to distinguish from other bayous along this reach that drain out of the Tunica Hills. Today, this part of the portage is marked by the interconnecting Lochlomand

and Babs Bayous. Threading through a bristly bottomland cane-brake, the bayou passage opened on to an oxbow lake rimmed with cypress trees. Known today as Lake Killarney, the French called this old Mississippi River cutoff Lac de la Croix (Lake of the Cross). Its placid water must have been a welcome relief to weary paddlers. At the north end of the lake, voyageurs nosed their canoes into the mouth of another bayou that led them out to the Mississippi riverbank. Modern maps label the slough tracing this old water-course Davis Bayou. Through its connection with the Mississippi in the eighteenth century, the depth of the watercourse rose and fell with the level of the river. When the Mississippi was high, travelers could traverse the portage in their canoes. At times of low water, voyageurs pulled their boats through the bayou passageways to and from the lake.[27]

Now enclosed within the walls of Angola Prison, one can still trace the bayou path of the Portage of the Cross through the penal farm's cultivated fields and cow pastures. Although the levee keeps the Mississippi out of the bayous, the old watercourse still drains the eighteen-thousand-acre penitentiary.[28] The Lake of the Cross, its distinctive fishhook shape discernible from the air, is slowly silting in from the east, making a natural transition from lake to cypress swamp. The penitentiary folks refer to the alligators here as "guards."[29]

◆ ◆ ◆

IN THE LATE 1600S, when the French began to explore the Lower Mississippi River, the Houma Indians occupied a bluff top vil-lage overlooking the portage. The location had obvious strategic and economic advantages. Early colonial leaders such as Henri de Tonti and Pierre LeMoyne d'Iberville regarded the Houmas, with an estimated force of 350 warriors, as essential allies for the fledg-ling Louisiana colony.[30] However, the region was in turmoil, and the Houmas did not remain at Old River for very long. Smallpox and other European diseases struck many of the native groups on the Mississippi, and slave-catching Indians allied with the Carolina English raided the smaller tribes relentlessly. In 1706, the Tunica Indians, hard pressed by Chickasaw slave raiders, fled their village on the Yazoo River and moved south. Instead of seeking out an unoccupied place to clear and establish their cornfields, the Tunicas commandeered the Houmas' village and fields, forcing the latter to relocate downriver. Like the Houmas, the Tunicas' stay at Old River was temporary; however, numerous landmarks in the area still bear their tribal name, including the loess hills around Angola.[31]

The colonial French were on the Mississippi River in increasing numbers during the early 1700s, and their use of the portage afforded the Tunicas the opportunity to carry on trade with the voyageurs and acquire coveted European merchandise, including muskets, blankets, iron and brass pots, and an assortment of other Old World accoutrements. The long-standing Tunica mission of the seminarian priest, Antoine Davion, also helped to endear the tribe to the French. With Father Davion's guidance, the Tunicas eagerly assimilated European material culture. When the Jesuit missionary, Pierre François Xavier de Charlevoix, visited the tribe in 1721, he found the chief dressed in French clothing.[32]

The Red River, sometimes labeled River Marne and Sablonnièr (Sandy) River on colonial French maps, positioned the Tunicas to expand their trading activities and reach Spanish settlements via a host of western tribes. Closest were the Avoyels, a small tribal group living about thirty miles up the Red River Valley, whose descendants are still in the Marksville area. In 1700, the French learned from the Avoyels about the great Red River Raft, a tangled mass of floating driftwood that clogged the channel, making canoe travel difficult and the passage of larger craft nearly impossible.[33] The raft's removal in the early nineteenth century is covered in Chapter 2.

◆ ◆ ◆

TWELVE YEARS after the Tunicas moved to the Portage of the Cross, French laborers began staking out a settlement at another

Indian portage far downriver. An Indian guide first pointed out the portage to Iberville during the French commander's initial exploratory voyage up the Mississippi. Iberville found that travelers could pull their canoes out of the river at this spot and drag them along a trail through the bottomland cypress to a stream (later named Bayou St. John), which flowed north into Lake Pontchartrain. The portage provided convenient access to the lake and a water route through Lake Borgne to French outposts on the Gulf Coast.[34]

The founding of New Orleans is one of those episodes in history loaded with consequences reaching far beyond the immediate goals at hand. While the riverside site at the Bayou St. John portage might be an excellent spot for a fishing camp, it has proven to be a perilous place for a major city. Protecting this vulnerable metropolis from the Mississippi River continues to shape US flood control strategy in the twenty-first century, a strategy in which Old River plays a critical role.

Jean-Baptiste LeMoyne, Sieur de Bienville, Iberville's younger brother, selected this unlikely location for the colony's new capital when he became governor general of Louisiana. Geographer Richard Campanella notes that Bienville's employer, the French Company of the West, needed a town close to the mouth of the Mississippi River to facilitate commerce and block English encroachment into the valley. Given the Company's objectives and the nature of the terrain, Bienville's choices were limited. The old portage site would have to do.[35]

The Crescent City's situation—located on a major river that is also an Atlantic shipping port—is certainly economically advantageous. However, New Orleans is within the floodplain of a powerful and capricious river. Riverside land may be high and dry one day and dissolved in the current the next. A lesser calamity, though just as bothersome, is the Mississippi's annual floods.[36] More than a decade after the city's founding, Edmé Gatien Salmon, the colony's commissary general, informed King Louis XV about the hardships of daily life beside the Mississippi: "[The location of New Orleans] in a flat and marshy country obliges all the inhabitants to dig little ditches in front of their houses, one or two feet in width by a foot or foot and a half in depth in order to drain off the water that seeps through the levee, when the river overflows."[37] To counter his critics, Bienville repeatedly emphasized the virtues of the place, citing close proximity to upriver plantations, the convenience of the Bayou St. John portage, and the good quality of the soil.[38] Indeed, the dirt on the river's shoulders, being sandy and well drained, is often the best soil around, and the banks of the active channel offer the highest ground in the floodplain; however, the governor ignored the settlement pattern of the indigenous inhabitants of the Lower Mississippi Valley. The Indians' riverside encampments were usually temporary accommodations for trading opportunities or fishing. Permanent dwellings were well removed from the active channel. Archaeological studies show that the native people of the region customarily settled in the uplands bordering the valley. Any village sites in the floodplain were usually on the natural levees of old cutoff channels. These oxbow lakes were calm, often a safe distance from the active river, and teeming with useful plants and wildlife.[39]

Despite the shortcomings of its location, the settlement of New Orleans gained momentum under the ambitious initiative of the Company of the West (later changed to the Company of the Indies). Although the Company employed some of France's top military engineers, making New Orleans habitable required years of perseverance. Then as now, those of higher social rank received lots on the high ground next to the river, while those less fortunate occupied the lower ground sloping down toward the swamp between the city and Lake Pontchartrain. As floodwaters swelled the Mississippi every spring, with severe inundations in 1719 and 1734, levees quickly became a familiar part of the New Orleans landscape. The first levees on the riverbank, only about six feet wide and three feet high, were soon washed away. The work of building and repairing these walls of dirt fell to the colony's enslaved African workforce. So vital were these earthen embankments that the Company routinely detained Africans arriving in the colony for New Orleans levee work before distributing the workers to upriver plantations. Likewise, slaveholders in the capital were obliged to contribute laborers to build and maintain the levee in front of the town.[40]

By the late 1720s, the Company and landowners along the Mississippi had thrown up a ragged chain of levees stretching thirty miles along both banks at New Orleans. Meanwhile, city residents found it necessary to build their own levees around individual houses. Annual flooding in the city streets seemed inevitable as the river overtopped the main levee; surged around it; or, more often, seeped under the poorly constructed pile of earth. The main levee at New Orleans in 1727 was still quite modest by modern standards, extending approximately a mile in length, from eighteen to twenty-two feet wide, and three feet high.[41]

In building their levees, the colonial settlers unwittingly revived an age old Mississippi Valley tradition. For thousands of years, the region's prehistoric societies dug, carried, and piled the earth's soil to construct ceremonial mounds. Scattered through the great valley, many of their monumental earthworks have endured centuries beyond the communities that made them. With the colonial French, mounding dirt began anew in the Lower Mississippi Valley and feverishly continues to this day. But the native people did not move the earth to control the river, though they certainly possessed the capability. For the prehistoric Indians, floods were not catastrophes. The annual rise and fall of the Mississippi River was part of an ancient and reassuring seasonal cycle. In sharp contrast to the modern mound builders described in the coming chapters, the Indians built their earthworks to influence the spiritual world, not the terrestrial one.

CHAPTER 2

Trouble the Water

Black powder explosions rocked the Old River area in January 1831, announcing the formidable presence of Henry Miller Shreve. A midlife portrait of the famous steamboat captain shows a man of action: barrel chest, broad shoulders, and lantern jaw, with flinty eyes beneath a high forehead.[1] Shreve was hewing a new channel for the Mississippi River across the neck of Turnbull Bend (see Figure 1.9). As the blasting smoke drifted away, a gang of laborers went to work with picks and shovels. While the men hacked at the loosened dirt to widen and deepen the cut, Shreve used the windlasses and cables of his steam-powered snag boat, *Heliopolis*, to drag scrapers filled with earth out of the trench. In the name of progress, the map of the Mississippi River was about to change, shortening the distance between Natchez and New Orleans by about eighteen miles.[2]

Shreve's cutoff transformed the active channel that looped around Turnbull Bend into a bracelet of slender lakes called upper Old River and lower Old River (see Figure 2.1). Turnbull Bend became Turnbull Island, and the Mississippi's confluence with the Red and Atchafalaya rivers was altered forever. The operation foreshadowed two busy centuries of modifications to the Mississippi River, reflecting an era of worldwide channeling and canal building on a grand scale.

The steam engine drove the Industrial Revolution of the eighteenth and early nineteenth centuries, making superhuman earth-moving projects a reality.[3] If engineering technology made it possible to straighten river channels, expediency rather than caution

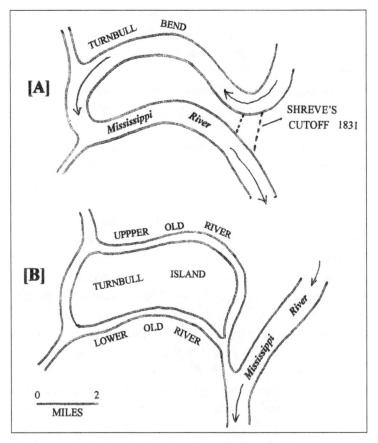

Figure 2.1 (A) Turnbull Bend prior to Shreve's cutoff, showing approximate location of the new channel, and (B) Turnbull Island with Upper and Lower Old River after Shreve's cutoff. Based upon Reuss, *Designing the Bayous*, 31.

often drove decisions about how to manage the country's waterways. Perhaps more than any other spot on the Lower Mississippi, the three rivers confluence was a place that warranted caution. In 1812 Major Amos Stoddard, who took possession of the Louisiana Purchase on behalf of the United States, recognized the Atchafalaya's potential to become the Mississippi's main channel. Historian Christopher Morris noted that colonial Spanish engineers and

administrators in the 1790s considered and then backed away from
an attempt to dig a cutoff in this vicinity—possibly across Turnbull Bend—fearing that the action might trigger a Mississippi River
diversion into the Atchafalaya Basin.[4]

By making his cutoff, Shreve disrupted a state of equilibrium
that had existed between the Mississippi, Red, and Atchafalaya Rivers since the sixteenth century. Turnbull Bend and the other two
horseshoe curves of this serpentine reach diffused the energy generated by a complex convergence of currents and sediment. In an
era of burgeoning river commerce, however, steaming around thirty-seven miles of twisting river bends to gain six miles of latitude
had become a tedious inconvenience. The Portage of the Cross, the
canoe passage exploited by Indians and French voyageurs of the
previous century, was useless in the steamboat era.

In addition to reducing travel time, Shreve believed that straightening out the Mississippi at Turnbull Bend would solve another
navigation problem. Shoaling on the Red River at a place called
Grand Point, near the mouth of the Red, sometimes made it difficult for steamboats to pass from the Mississippi into the valleys
of the Red and Ouachita.[5] Shoaling, riverboat jargon for shallows,
occurred where a slow-moving current released enough sediment
to raise the riverbed. With his cutoff, Shreve thought that he could
speed up the current through the Grand Point reach by triggering a
sort of hydraulic chain reaction. The cutoff would shorten the channel distance between the Mississippi's elevation above and below
Turnbull Bend from eighteen miles to just a few hundred feet. At
the new channel cut, the Mississippi's surface would fall enough to
give the mouth of the Red River a steeper gradient. The riverboat
pilots who plied the Red and Ouachita hoped the improved gradient would accelerate the Red River's current enough to clear out the
bothersome shoals. Many also believed that removing the Mississippi's meanders with cutoffs would eliminate crevasses by relieving
the pressure on outside bends. What really happened in the aftermath of Shreve's cutoff prompted Louisiana's first state engineer,

George T. Dunbar Jr., to observe in hindsight that "the effect of an extensive cut-off, in so large a stream as the Mississippi, I am confident, is not generally understood."[6]

◆ ◆ ◆

INDIAN REMOVAL, the Louisiana Purchase, emergent sugarcane and cotton plantations, and the advent of steamboats all played a part in transforming the Old River region during the half century leading up to Shreve's cutoff. Significant demographic changes began after the Treaty of Paris in 1763, marking the conclusion of the Seven Years War, better known in North America as the French and Indian War. Although that conflict's battles raged elsewhere, the war had profound repercussions in the Lower Mississippi Valley and Atchafalaya Basin. France ceded New Orleans and Louisiana west of Mississippi to the Spanish before the war's end, while the Treaty of Paris awarded French Louisiana east of Mississippi River to the English.[7]

In the wake of the treaty, the political landscape for many of the Indian tribes living east of the Mississippi abruptly shifted. The Tunicas and several other small tribes in the region, including the Biloxis, Pascagoulas, Houmas, and Ofogoulas, had been longtime allies of the French and thus enemies of the English. Since these Lower Mississippi Valley Indians did not fight in the war, they naturally resisted its consequences. In early March 1764, some of the warriors from these pro-French tribes ambushed a company of British soldiers led by Major Arthur Loftus, traveling up the Mississippi River from New Orleans to occupy Fort Chartres in the Illinois Country. As the expedition's longboats passed the mouth of the Red River, rounded Turnbull Bend, and approached the heights on the east bank known as Davion's Rock (named for the longtime missionary to the Tunicas), the Indians fired at the redcoats from positions on both sides of the Mississippi. To his disgrace, Major Loftus abandoned his mission, ordered his boats turned about, and had his force of over three hundred men paddle hastily back to New Orleans.[8]

Soon afterward, locals in the area began calling the bluffs at Davion's Rock Loftus Heights to England's embarrassment. For the Tunicas and the other pro-French tribes in the area, the Loftus ambush
was their last swipe at their old enemy. During the 1770s, these Indians continued to maintain villages in the Pointe Coupée-Old River
area, finding employment in various capacities among the local
plantation communities. Maps of the region show that these groups
remained here until the 1780s, when Spanish land grants began to
overlap their settlements. At that time, colonial Spanish authorities
invited the tribes to move up the Red River to new village sites on
the Marksville Prairie.[9]

The French and Indian War immediately preceded the arrival
of the Acadian people in the Atchafalaya Basin. In a forced migration not unlike that experienced by the Choctaws, Creeks, and
other tribes in the early nineteenth century, British encroachment
into Nova Scotia and New Brunswick pushed the Acadians, in
their colonial French communities, from their adopted homeland.
During the war they dispersed aboard British ships, some to France
and others scattered among Britain's American colonies. Between
1765 and 1785, a number of these exiled groups reunited in Spanish
Louisiana, forming Acadian settlements along the Mississippi River
near the Bayou Lafourche outlet and in the western Atchafalaya
Basin along Bayou Teche.[10]

Though officially under Spanish control, Louisiana's population
remained predominantly French. The late-eighteenth-century plantations on west side of the Mississippi, arrayed along the great meander bend at Old River, were an outgrowth of the colonial settlement
at Pointe Coupée, some twenty miles downriver. This French enclave
dated to the colonization efforts of the Company of the West, beginning in the early 1720s. Planters here, exporting tobacco, indigo,
and lumber, introduced the Old River region's largest ethnic group,
free and enslaved Africans. The geographical isolation of the Pointe
Coupée area shaped what historian Gwendolyn Midlo Hall calls a
"face-to-face community" of Africans and Europeans speaking Louisiana Creole, an amalgam of French and several African languages.[11]

On the east side of the river, 1770s maps of the Old River area reveal a string of English land grants crowding the waterfront. But the maps are misleading; most of the grantees never took possession of these properties. More lasting residencies followed Spanish control of the region during the American Revolution, which resulted in a new round of land grants to soldiers and government officials. After Spain relinquished her Lower Mississippi Valley holdings, the US government recognized the Spanish grants. Maps and land records indicate that English, Scottish, and Irish tenants began acquiring and settling these properties during the first decades of the nineteenth century. These families from the failing tobacco plantations in the Chesapeake region brought with them English-speaking African American slaves, the vanguard of the nearly one million enslaved people eventually transferred into the Deep South to labor in the cotton and sugarcane fields.[12]

To the diverse population coalescing in the Old River region, the specter of disaster rode the river's annual rise. Although the Mississippi usually kept within its banks, little could be done to fend off the devastating floods that rolled over the riverside plantations in 1809 and 1825.[13] Yet the promise of wealth outweighed the risks now that the waterway powered the young country's economic and territorial expansion. From the 1780s, flatboats and keelboats stuffed with trans-Appalachian commodities rode the currents down the Ohio River to Cairo and on to Natchez and New Orleans by the thousands. After 1812, steamboats gradually transformed the river into a two-way superhighway for commerce.[14] But the Mississippi and its tributaries were rife with hazards, and by the 1820s Congress recognized a federal responsibility to improve the navigation of the western rivers. The task fell to Henry Shreve.

◆ ◆ ◆

SHREVE WASN'T THE FIRST to attempt a cutoff at Turnbull Bend; a local landowner, Francis Routh, had already tried to cut a channel across the point. Routh owned an extensive riverfront tract on the east side of the Mississippi, a portion of which later became Angola plantation, the eventual site of Louisiana's state penitentiary.[15] In 1829, Routh used his slaves to dig a trench across the narrow neck of the bend that was deep enough to capture part of the Mississippi's flow during high water but too shallow to effect a channel diversion.[16]

Depending upon the river's dynamics at any given spot along its length, artificial channel diversions are never a sure thing. The composition of the underlying sediment and gradient characteristics of the potential new channel compared to that of the existing riverbed help determine whether or not the river will adopt a new course. Famously, General Ulysses S. Grant tried without success to reroute the Mississippi away from Vicksburg in 1863 by digging a channel across De Soto Point. Ironically, the river diverted naturally through Grant's cut across the point a decade after the end of the Civil War.[17]

Shreve had also experienced the Mississippi's recalcitrance. Before digging his channel across Turnbull Bend, Shreve's outfit cut a channel through Bunch's Bend, about 150 miles to the north near Lake Providence, Louisiana.[18] Unlike Shreve's Turnbull Bend cutoff, which took the river's full flow in a matter of weeks, the Bunch's Bend cut proved to be a slow conversion, taking more than a year to capture the Mississippi.[19]

Routh's private initiative indicated that there was local interest in making a cut across Turnbull Bend, and the job must have appealed to Shreve's love of a challenge. A larger-than-life hero living in a century crowded with heroes, Shreve had helped Andrew Jackson win the Battle of New Orleans and scuttled Robert Fulton's Mississippi River steamboat monopoly.[20] His voyages up and down the Mississippi over the previous thirty years had taken him around Turnbull Bend many times, first aboard flatboats and keelboats, and later at the helm of steamboats of his own design. During the long hours spent churning upstream around the bend's eighteen-mile circuit, the idea of digging a cutoff had undoubtedly crossed his mind. According to an 1843 report by George T. Dunbar, the narrowest part of Turnbull Bend was only 160 feet across; however, the problem of assembling an army of diggers to move thousands of tons of earth must have seemed insurmountable.[21]

When Shreve returned to the Turnbull Bend reach in 1830–1831, he brought the solution to the problem: a workhorse of a riverboat named *Heliopolis*. Although *Heliopolis* proved a most capable channel digger, she was designed for another purpose. The cutoff at Turnbull Bend was only a sideshow alongside Shreve's official mission to the Lower Mississippi Valley that year.

In the early nineteenth century, the US Congress considered snags, uprooted trees lodged in the channels of the Mississippi and its tributaries, the chief impediment to western trade. This dangerous driftwood was ubiquitous in the Mississippi river system. Along hundreds of miles of forested river bank, an endless cavalcade of trees slid into the water as the streams meandered and flooded. While the trunks were floating, they presented a serious hazard, but as they became waterlogged, the heavy root end typically sank to the bottom, standing the tree up vertically in the current and making them still more dangerous. In this position, snags became deadly traps capable of capsizing flat boats and piercing the hulls of steamboats. River folk recognized two kinds of snags: sawyers and planters. A congressional committee in 1830, charged with addressing this troublesome issue, provided an eloquent description:

> At many places the river rushes with immense impetuosity through whole forests of these trees, the tops of which are partly seen above the surface of the water, which gives them a motion accompanied by a noise not unlike that made by the motion and noise of a saw mill; hence the name "sawyer" which has been given to them by the boatmen. Others of those trees lie concealed below the surface of the water, the ends projecting up or down the river as chance may place them; they give neither motion nor noise, and, from their stationary position are called "planters."[22]

To rid the western waterways of this menace, Shreve developed a steam-powered snag boat that was a combination battering ram and floating sawmill. This extraordinary vessel employed two separate steamboat hulls, each with its own boiler, smokestack, and paddlewheel, joined at the bow by an iron-plated wooden beam more than two feet thick. The business of pulling a snag commenced with the pilot taking aim and steaming toward his target. When the bow's horizontal beam, skimming just above the water line, plowed into a snag, even the largest and most obstinate trunks could not withstand the force brought to bear by the twin paddlewheels. The collision either wrenched the snag out of the river bottom sediment or broke it off above the root ball. Once dislodged, the crew secured the trunk with ropes, and the vessel's powerful windlasses, driven by the two steam engines, hauled the catch across the bow beam and down between the hulls where the crew sawed the tree into harmless chunks. Shreve's specialized boats easily handled snags as long as 160 feet and the whole operation, from initial impact to final saw cut, took less than an hour.[23]

Shreve conducted his snag removal project under the auspices of the fledgling US Army Corps of Engineers, brandishing the title of superintendent of western river improvements during the presidential administrations of John Quincy Adams and Andrew Jackson.[24] Between October 1830 and March 1831, Shreve's boat *Heliopolis* removed 1,334 snags on the Mississippi River below the mouth of the Ohio. Along the way, his crews cut trees from over five hundred miles of riverbank to stem the tide of timber caving off into the water.[25] Shreve's success earned him tremendous popular acclaim. In 1830, the *New York Herald* trumpeted that the Mississippi River had been rendered "as harmless as a millpond."[26] Such optimism, however, was premature. In truth, the river remained a dangerous waterway. Despite Shreve's efforts, snags continued to plague river traffic. Out of sixteen steamboats that sank in the Old River reach between 1830 and 1912, seven went down because of collisions with snags.[27]

It was during a break from his snag-pulling work that Shreve turned his attention to Turnbull Bend. The crew of the *Heliopolis*, possibly augmented by slave labor from Routh's plantation, began by clearing the trees away from the path of the channel in early January 1831.[28] The place where they worked was no deserted stretch of river bottom. Shreve's cut severed an important road that ran the length of the point, out to a Mississippi River ferry landing at the tip of Turnbull Bend. The west bank ferry landing was situated midway between the mouth of the Red River to the north and the outlet for the Atchafalaya distributary on the south. The ferry, the only conveyance across the river between Natchez and St. Francisville, linked overland travelers with the Mississippi towns of Fort Adams and Woodville and the Louisiana towns of St. Francisville and Alexandria.[29]

Any wayfarers on the Turnbull Bend road that January would have had to seek an alternate route after Shreve's blasting powder began to enlarge Routh's ditch. Beginning at the southern end of the cut and moving steadily up the trench toward the upstream end, the workers sunk casks filled with the explosives down into shafts dug to depths of around twenty feet. Once the explosions loosened the ground, the real challenge was getting the dirt, close to twenty thousand tons of it, out of the trench. To handle this colossal task, Shreve positioned *Heliopolis* at the southern end of the cut, where the new channel would flow out into the Mississippi (see Figure 2.1).[30]

With the snag boat's bow nosed in against the shore, the big windlasses that normally reeled giant logs up from the river bottom were deployed as draglines to pull two large scrapers, open-ended iron or wooden boxes, which operated nonstop clearing out the sandy dirt dislodged by the explosions. One set of cables stretched out into the trench and looped around return sheaves, vertical posts that functioned as pulleys. These towlines hauled the scrapers out to the blast area, and a second set of cables pulled the loaded scrapers back toward the boat to be emptied into the river. As the channel excavation progressed northward, Shreve's men repositioned the return sheaves at the advancing head of the cut. For two weeks, Shreve's crew stoked the fireboxes of *Heliopolis's* twin boilers, black plumes rolled from the smokestacks, the windlasses whirred, and the scrapers dumped ton after ton of earth into the Mississippi's current.[31]

Shreve's excavation progressed rapidly thanks to the nature of the sediments comprising the narrow neck of land. The point bar that shaped Turnbull Bend was composed of a series of broad, parallel, sandy ridges with vertical thicknesses between forty and one hundred feet. Between the low ridges were shallow swales filled with silt and clay. Aligned north to south, the ridges ran in the same direction as Shreve's cut and the sandy soil yielded readily to the blasting powder and scrapers.[32]

On January 28, 1831, just fourteen days after the digging started, the Mississippi began flowing down Shreve's cutoff. Two days later, *Heliopolis* and the steamboat *Belvidere* made successful runs upstream through the enlarging gap. Shreve's initial cut was only seventeen feet wide and twenty-two feet deep, but the neck of Turnbull Bend dissolved into the Mississippi's torrent like sugar in coffee. According to the New York newspaper, *The Evening Post*, "in five days [the cutoff] was the main channel of the river, being half a mile wide and of equal depth with the other parts of the river."[33]

The speed with which the Mississippi accepted this radical diversion says much about the volatile nature of the Old River area. In addition to the sand-laden point bars built by the river's more recent meanderings, deeper beds of erodible sand mark the buried meander belts of the Mississippi's ancient channels through this reach. Of course, Shreve could not have known what lay beneath the sandbars and cutbanks; these secrets awaited the deep coring explorations of the twentieth century. In 1831, there was much to learn about the geology and hydrology of the Lower Mississippi Valley. An education in the hazards of making artificial changes to the Mississippi's channel came swiftly to west bank landowners opposite the mouth of Shreve's cutoff. The river's current surging out of the new channel collided head-on with the opposite riverbank, eventually caving the land back three-quarters of a mile.[34]

◆ ◆ ◆

WHEN SHREVE PILOTED *HELIOPOLIS* back up the Mississippi in early 1831, he left behind more than a severed river bend. The cutoff that came to bear his name—and dampen his reputation— stirred a whirlpool of dilemmas at Old River. The most serious repercussion, the cutoff's role in influencing the diversion of the Mississippi's channel, uncoiled gradually over the course of a century as the three rivers adjusted to the new regime; however, one consequence manifested itself practically overnight. Bereft of the Mississippi River, the water level in Turnbull Bend fell two-and-a-half feet in thirty-six hours.[35]

The increased fall gave the Red a slightly steeper gradient and, true to predictions, the accelerated current through Grand Point was sufficient to alleviate the shoaling problem. But nature often plays a zero-sum game. Shreve's cutoff abruptly shifted the Mississippi's active channel six miles away to the east. The Red River, carrying barely one-eighth of the Mississippi's volume, emptied into the abandoned river bend and lost its momentum in the stagnant upper arm of Old River (see Figure 2.1). With its slowed-down current, the Red River was soon redistributing the sediment from the Grand Point shoals to the new intersection with the Mississippi. Seven years later, Shreve was back at his channel cut, dredging the Old River shoals in response to a chorus of complaints from riverboat pilots, perhaps the same ones who had clamored for the cutoff to clear Grand Point.[36]

The other river at the junction, the Atchafalaya, also reacted swiftly to Shreve's cutoff. No longer positioned to capture its share of the current rolling around Turnbull Bend, the Mississippi's largest distributary dwindled to a trickle. And like the upper arm, Old River's lower arm began to constrict as it filled with silt. We don't know how Shreve felt about cutting off the distributary or if he even considered this consequence. The Atchafalaya may have been of little concern because of the ancient raft of timber lodged about thirty miles below the river's outlet from the Mississippi. The impassable jumble of jammed driftwood, which extended some forty miles downriver, had always prevented boats from using the northern end of the waterway as a transportation route. Although the upper Atchafalaya had been useless for boat traffic, the water that flowed around and beneath the raft was vital for the communities and planters down below in the Atchafalaya Basin. The Lower Atchafalaya River was the heart of a busy steamboat network linked to the Grand River and bayous Courtableau, Teche, Maringouin, Grosse Tete, Lafourche, and Plaquemine. Shreve's cutoff could not be reversed, but residents in the Atchafalaya Basin began to wonder if clearing the raft might open the way for water from the Mississippi River to resume flowing into the distributary.[37]

◆ ◆ ◆

IF SHREVE'S CUTOFF TIPPED the Mississippi's inclination toward a new path to the Gulf, it was the removal of the Atchafalaya raft that released the brake. In the early days of river commerce, rafts were a common nuisance in Louisiana's rivers and bayous.[38] During periods of flooding and riverbank caving, rafts collected floating trees on the upstream end faster than they lost the old, disintegrating timber on the downstream end. Best known was the Great Raft on the Red River, over 160 miles of snarled trees lodged together so tightly that one could cross the river dry shod. Shreve and his snag boats first cleared the Red River raft between 1833 and 1838. In gratitude, the residents at the raft's upper end immortalized the steamboat captain by naming their new settlement Shreve's Town (later changed to Shreveport). However, the Red River proved the celebration a bit premature; within a year, the Great Raft had reformed and closed the river. Clearing work ensued, although the problem persisted for years to come.[39]

While Shreve's federally funded snag boats were working to open the Red River to steamboat passage, the responsibility for clearing the Atchafalaya raft fell to the people of Louisiana. In the early days of settlement in Attakapas country, as the western Atchafalaya Basin was sometimes known,[40] the blocked artery was a tolerable inconvenience. Unfettered access through the Upper Atchafalaya River became a larger issue after the Louisiana Purchase and the spread of cotton and sugarcane plantations throughout the Attakapas region. A century earlier, when the villages of the Houmas and Tunicas guarded the Portage of the Cross, the three rivers area had been a busy intersection for dugout canoes (which could weave their way through and around the rafts). By the 1830s, if the rafts could be cleared, the Old River junction stood to become a similar hub for riverboat commerce, with spokes jutting into the Atchafalaya Basin, the valleys of the Red and Ouachita Rivers, and the Mississippi's upper and lower valleys.[41]

Official efforts by the state of Louisiana to clear the Atchafalaya raft finally began in 1839. The previous year, the Corps of Engineers reported that the raft's removal would significantly improve the Atchafalaya's ability to draw water from the Mississippi, thus relieving the slowdown caused by Shreve's cutoff. If the federal stamp of approval lacked a corresponding congressional appropriation, the report spurred the Louisiana legislature to launch an attack on the raft through the state's newly created Office of Public Works. Initially, Louisiana's makeshift snag boats and their crews proved to be ineffective. Local landowners even lent a hand by assaulting the raft with fire. Except for "roasting a number of alligators," as historian Martin Reuss noted, the flames inflicted little damage to the ancient tangle of interlocked trees.[42]

By the early 1840s, perseverance began to pay off, with measureable headway coming after the legislature appointed the first state engineer, George T. Dunbar, in 1842. Professional direction by Dunbar and his successors eventually cleared the Atchafalaya raft, but the difficult operation proceeded slowly. As Louisiana's work crews and snag boats removed more and more of the raft, the distributary's flow increased and its channel grew deeper and wider. The Atchafalaya was indeed confirming the Corps of Engineers' prediction, but not in the way that the engineers envisioned. The increasing flow rushing down the Atchafalaya wasn't coming from the Mississippi; the Atchafalaya was stealing the Red River.[43]

◆ ◆ ◆

THE NEW ORLEANS *TIMES-PICAYUNE* CALLED IT "A NEW NOTION."[44] But many riverboat captains and engineers had long recognized something ominous about the way the rivers were reconfiguring themselves at Old River. In his 1847 report to the Louisiana legislature, state engineer Paul O. Hébert put the situation in stark terms: "[I] am firmly of the opinion that the Red River and Atchafalaya were one and the same river. The Red River has now a tendency, hastened by the effects of 'Shreve's cut-off,' to return to the Atchafalaya; and this will be consummated in a few short years, unless prevented by extraordinary measures."[45]

Two powerful political constituencies viewed this development with alarm. Cotton planters in the Red and Ouachita valleys faced losing their shipping route down the Mississippi to New Orleans. If Old River were to close completely and the Red divorce itself from the Mississippi, cotton growers in the Louisiana parishes of Avoyelles, Catahoula, LaSalle, Rapides, Grant, and Natchitoches, along with more northern communities, would have to get their product out to the world through the Atchafalaya Basin. Two options had feasibility; both were unreliable and would involve increased costs.

During periods of high water, steamboats might be able to enter the Mississippi and reach New Orleans through Bayou Plaquemine.

The other outlet was to reach the Gulf of Mexico via the Lower Atchafalaya River. The real problem with both options was the Atchafalaya raft. Although parts of the raft had been cleared by the 1840s, large sections still blocked boat traffic. Until the raft was completely cleared, the planters on the Red and Ouachita systems faced the possibility of being landlocked. In an era when passable roads from the east barely reached the Mississippi Valley and the nearest railway system linked to a seaport was over six hundred miles away, the prospect of losing a water transport route was untenable.

The other political constituency, the one that held the most power, was New Orleans. Importantly, the Crescent City hosted the annual sessions of the Louisiana legislature until the end of the 1840s. As the main shipping port for the entire Mississippi River Valley, money flooded into the city across a waterfront nine miles long bustling with seagoing merchant ships and riverboats laden with bales of cotton and sugar by the hogshead. If Old River closed and the Red and Atchafalaya formed one river down to the Gulf, the cotton producers of the Red River and Ouachita River valleys might be obliged to bypass New Orleans, realigning commerce in Louisiana to the detriment of the state's wealthiest city.[46] As Hébert indicated, the situation demanded bold initiative, and a few politicians and seasoned steamboat captains thought they saw a way to clear the Old River shoals and keep the vital passageway open. The proposed solution was a controversial one that called for digging another Mississippi channel cutoff, this one across the neck of Raccourci Bend.[47]

Riverboat pilots began ringing the warning bells in the fall of 1838; sediment in the Old River channels threatened to deny them passage from the Mississippi to the Red River. Without swift action, the season's cotton crop ripening on the plantations up the Red and Ouachita might not be able to reach the market in New Orleans. Many blamed Shreve's cutoff for the nuisance that was becoming a matter of grave concern for the New Orleans business community and for the Red and Ouachita planters. The federal government had

funded Shreve's operation and now the federal government must set
things right. Accordingly, the New Orleans Chamber of Commerce
fired off a letter to the US War Department. Within weeks, Captain
Shreve was packing his travel bag in St. Louis, while a Corps of Engineers dredge boat stationed at the mouth of the Mississippi began
chugging upstream. Shreve and the dredge converged on Old River
in time to clear a pass through the shoals, but all concerned viewed
the fix as temporary. Some embraced a more proactive solution.[48]

While the dredge boat fought with the Old River sediment, the
word *Raccourci* buzzed through the state legislature convening
in New Orleans. A map depicting the Old River area in the 1770s
shows a small community by that name perched at the south end
of the meander loop that came to be known as Raccourci Bend.
Here, the river's channel below Turnbull Bend formed a great "S"
curve, running south to Raccourci, turning and flowing back to the
north for about six miles before rounding Tunica Bend (passing the
lower entrance to the Portage of the Cross) and finally streaming
south again.[49] Raccourci being French for shortcut, the settlement
may have gotten its name from the short distance across the neck of
Tunica Bend. The 1770s map shows a village of Ofo Indians living
beside the river on the other side of the neck (see Figure 1.9).[50] By
the early 1800s, both Raccourci and the Ofos were gone to make
way for Spanish land grants.

A channel across the neck of Raccourci Bend would divert the
Mississippi coming through Shreve's cutoff toward the southeast, to
the top of Tunica Bend, and then south down the existing channel
along the base of the Tunica Hills (see Figure 2.2). The operation
would save approximately nineteen miles of river travel, but like
Shreve's cutoff, the purpose behind making a cutoff across the neck
of Raccourci Bend was about more than saving time.[51] Ironically,
the same arguments floated forth in favor of Shreve's cutoff—the
need to clear the Grand Point shoals—now circled back to support
the Raccourci cutoff. Raccourci proponents argued that the new cut
would lower the water level of the Mississippi, once again creating

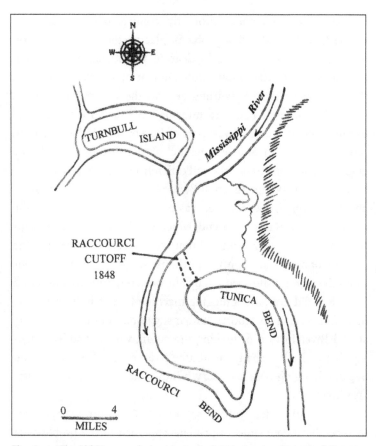

Figure 2.2 The Old River area between 1831 and 1848, showing the path of the Raccourci cutoff. Based upon Norman's Chart of the Lower Mississippi River by A. Persac, 1858.

a fall that would speed up the Red River's current enough to clear out the Old River shoals. This, in turn, would (proponents hoped) drain the area above Old River more efficiently, reducing the danger of flooding on the Lower Red and Ouachita.[52]

In January 1843, increasing talk in the Louisiana legislature about making the Raccourci cutoff prompted the members of the house to pass a resolution calling for a report on the issue by the state

engineer. George Dunbar promptly confirmed the seriousness
of the Old River shoaling problem, placing the blame partly on
Shreve's cutoff and partly on the clearing of the Atchafalaya raft. On
the subject of the Raccourci cutoff, Dunbar urged caution. Beyond
the unknown elements swirling around the Raccourci proposal,
he gave the house some hard information. Dunbar estimated the
distance across the narrowest part of Raccourci Bend at one mile,
much farther than Shreve's channel across Turnbull Bend. The
engineer also pointed out the soil differences between the two cutoff operations. Where Turnbull Bend was a sandy point bar favorable to creating a new channel, clay deposits comprised half of the
distance across Raccourci Bend, which would resist the Mississippi's scouring. Dunbar reckoned that the project would require the
removal of more than eighty-four thousand tons of earth, over four
times the amount shoveled out by Shreve's crew and *Heliopolis*, and
Louisiana did not have a digging machine to match Shreve's snag
boat. Concluding his report, Dunbar suggested continued dredging
at Old River instead of making the cutoff, with the added warning that a channel change would be irreversible. If further action
beyond dredging were necessary, he advocated closing the Atchafalaya River.[53]

Two months after Dunbar submitted his report, a flood submerged the state engineer's prudent approach to the problems at
Old River. In March 1843, rising water inundated the Red River bot-
tomlands of Rapides Parish below Alexandria and spread over large
areas in the Ouachita drainage of Catahoula Parish. Planters with
fields drowned beneath three feet of water saw the Raccourci cutoff
as their salvation and lost no time in communicating their position
on the issue to the legislature. Despite mounting support among
lawmakers in favor of the cutoff, opponents managed to table the
subject until the next session.[54]

In 1844, proponents of the Raccourci cutoff finally got their victory, but just barely. The house was almost evenly divided on the
issue, and the New Orleans *Times-Picayune* reported "considerable

debate." On March 11, the house narrowly passed the resolution
authorizing the cutoff, while the senate at first rejected the resolution (by one vote) and finally passed it on March 23. Governor Alex-
andre Mouton signed Resolution No. 69 into law on March 25, 1844,
instructing state engineer Dunbar to "repair to the Raccourci Bend
. . . with such number of state hands as may be necessary, together
with any machinery belonging to the state he may deem requisite,
and immediately cause the Cutoff." Dunbar undoubtedly considered the resolution's call for a swift denouement rhetorical instead
of realistic; the work on the cutoff would end up dragging out over
four long years. Since the resolution did not include funding, the
governor and legislature apparently intended for the project's cost
to be paid out of the state engineer's annual appropriation.[55]

The "state hands" at Dunbar's disposal were inmates of Louisiana's penitentiary, located in Baton Rouge. Work on the Raccourci
cutoff happened to coincide with a radical change in the state's
penal system, a move calculated to save the state treasury close to
half a million dollars per year. In 1844, the legislature turned the
penitentiary operation over to a private entity, McHatton, Pratt and
Company, which paid nothing for the privilege of farming out the
inmates for profit to contractors and plantation owners. In effect,
the law enslaved the state's convicts. The Louisiana treasury even
contributed toward the company's expenses to clothe and feed the
prisoners.[56]

Work was finally underway in November 1844, with state hands
among the five hundred laborers clearing the path for the new
channel. In an effort to assist the underfunded state engineer's
office, proponents of the Raccourci cutoff solicited contributions
from planters in the parishes affected by high water in the spring of
1843. Not long after work commenced, the New Orleans Commercial Bulletin deemed the fund-raising committee "so far very successful." By December, however, meeting expenses for the workers'
food and shelter proved a challenge. Backers of the project reached
out beyond the immediate area to landowners in Concordia and

Tensas parishes trying to raise an additional $3,000. Appeals for cash also went out to steamboat operators, who were told that the shortcut would save them $25 in firewood expenses. Meanwhile, anxious planters hoped the cutoff would be completed before the spring season high water on the Red, Ouachita, Tensas, and Black Rivers. If any of these "investors" in the project happened to visit the Raccourci cutoff work site in the spring of 1845, they found the place deserted.[57]

◆ ◆ ◆

PAUL OCTAVE HÉBERT assumed the office of state engineer on April 1, 1845. A future governor of Louisiana, he would also serve as a brigadier general in the Confederate army, in command of a base at Galveston, Texas. Historian Thomas W. Cutrer notes that Hébert's men saw him as "aristocratic and imperious."[58] Hébert found his New Orleans office swamped with serious navigation problems on Louisiana's many bayous and rivers, and the Raccourci project was temporarily shelved. When he managed the time to inspect the abandoned works at the Raccourci cutoff, Hébert termed the project "far from finished." The previous year, the *New Orleans Commercial Bulletin* reported the length of the proposed channel cut to be less than three-quarters of a mile, much reduced from Dunbar's one mile estimate. After Hébert's surveyors made an accurate study, the state engineer set the official length of the cut at seven-eighths of a mile.[59]

Like his predecessor, Hébert was strongly opposed to the Raccourci cutoff. His annual report for 1845 led off with a progress report on navigational improvements, with Raccourci going unmentioned. Hébert's narrative then segued into a manifesto on the folly of making cutoffs on the Mississippi River. In vivid terms, he predicted "serious injury" for those below the Raccourci cutoff, on Tunica Bend and further south at Pointe Coupée. To back up his argument, Hébert noted that planters on Raccourci Bend below Shreve's cutoff *wanted* the Raccourci cutoff for this very reason. After the 1831 cut-off, the redirected Mississippi washed away miles of riverfront property and continued to gnaw at the valuable cotton land with every spring rise.[60] The Raccourci cutoff would take the active river away from those beleaguered planters on Raccourci Bend.

Even pro-Raccourci legislators at loggerheads with Hébert's stand on the cutoff issue could not argue with the state engineer on his next point, namely, that the federal government should maintain the Mississippi and its tributaries, not the states that border the great rivers. Emphasizing more far-reaching responsibility, Hébert referred to the Mississippi as an "inland sea" or "ocean river," asking, "Why should the Treasury of the State of Louisiana be drained for works, the benefit or bad effects of which, as the case may be, will be felt by ten other States of this Union?"[61]

With prescience borne out eighty-one years later by bitter experience, the state engineer also stressed the need for outlets along the Lower Mississippi River, instead of more levees:

Every day, levees are extended higher and higher up the river—natural outlets closed—and every day the danger to the city of New Orleans and to all the lower country is increased. Who can calculate the loss by an overflow to the city of New Orleans alone? Instead, therefore, of throwing suddenly a larger quantity of water into the Lower Mississippi and elevating its level by opening cut-offs above, we should, on the contrary, endeavor to reduce this level, already too high and too dangerous, by opening all the outlets of this river. We are every year confining this immense river closer and closer in its own bed—forgetting that it is fed by over 1500 streams—and regardless of a danger becoming every year more and more impending.[62]

To landowners suffering from flooding on the Lower Red and Ouachita, he recommended, as an alternative to making the Raccourci cutoff, the clearing and improving of the Atchafalaya River so that it could carry away the Red River's annual rise. The improved Atchafalaya would also serve as a much-needed outlet to relieve high water on the Mississippi. Tactfully, Hébert assured that he would, if ordered by the legislature, "cheerfully" carry the Raccourci cutoff through to completion.[63]

♦ ♦ ♦

THE ATMOSPHERE IN THE LOUISIANA GENERAL ASSEM-BLY was anything but cheerful in March 1846, when legislators wrestled over the Raccourci cutoff. Committees formed on both sides of the issue and trotted out expert witnesses to support their stands. What is extraordinary about the ensuing debate is the "apples vs. oranges" nature of the two groups of opposing witnesses. The committee promoting the Raccourci cutoff relied solely upon the testimony of local residents and steamboat captains, men who knew the Mississippi and Red Rivers intimately. On the other side, the committee seeking to block the cutoff stood behind depositions from a panel of engineers, most of whom were associated with the Louisiana Office of Public Works, including current and former state engineers. The two approaches to the issue, both equally cred-ible, were as different as chalk and cheese.

The witnesses in favor of the Raccourci cutoff were Walter Turn-bull (local landowner), William M. Wilson (steamboat captain), Hiram Wilson (steamboat captain), John M. Phillips (former local resident and steamboat captain), Peter Dalman (steamboat cap-tain), David Maguire (steamboat captain), and J. J. A. Plauché (long-time resident familiar with the Red River). With little variation, these river men testified that Old River was silting closed, cutting off the passage between the Red and the Mississippi. Their argu-ment was much the same as that for Shreve's cutoff. The solution, they all chorused, lay in reducing the height of the Mississippi at its confluence with the Red. The cutoff would accomplish this goal by lowering the Mississippi between two and three feet (the difference in elevation between the upper end of the cut and the lower end). In response to the added vertical distance or fall between the Red and the Mississippi, the Red would flow faster through Old River, clearing out the shoals. Several deponents referred to Shreve's cutoff and the subsequent clearing of the Grand Point shoals (without mention of the Old River shoaling caused by the 1831 cutoff). Most of the river men foresaw no dangers or damage for those living downriver of the cutoff.[64]

On the other side of the issue, the engineers agreed, but disagreed. They were, along with Hébert, Claudius Crozet (former Louisiana state engineer), George T. Dunbar (former Louisiana state engineer), Major William S. Campbell (civil engineer and member of the Louisiana legislature), T. B. Leonard (USACE civil engineer), and Major Harry Joseph Ranny (civil engineer based in New Orleans). These men testified that the cutoff would indeed lower the height of the water at the Mississippi's confluence with the Red, but argued that the result would not be as the river men envisioned. Instead of clearing the shoals, the increased current through Old River, combined with the shallower depth, would make steamboat passage even more difficult, especially during periods of low water. However, the worst effects, the engineers agreed, would be visited upon the people downstream along Tunica Bend and at Pointe Coupée. T. B. Leonard warned of "most mischievous consequences."[65]

The problem with the expert testimony on both sides wasn't whether one group had more insight into the question than the other. They all knew the Mississippi and the Red. What they all lacked was experience with artificial cutoffs on the Lower Mississippi River. Shreve's cutoff provided the only example (unless some of them were familiar with Shreve's upriver Bunch's Bend cutoff). For both the river men and the engineers, one or even two case studies hardly allowed for accurate predictions about what might happen following swift and radical redirection of a river such as the Mississippi.

Apparently bewildered by the hyperbole and hydrologic speculation on both sides, the 1846 Louisiana legislature chose to do nothing by rejecting a bill to fund the Raccourci cutoff.[66] Throughout the remainder of the year, the state engineer's men and resources deployed to the Teche, Lafourche, Tensas, and several other streams, struggling to clear miles of shoals, snags, and rafts. Due to a lack of funds for dredging equipment, riverboats continued to drag through the shoals in the Old River passage between the Mississippi and the Red during times of low water. Frustrated with the overwhelming challenges to keeping Louisiana's waterways navigable, Hébert shared with the legislature a bit of wishful thinking: "As a remedy I would suggest a resort to a system of Rail Roads. Well constructed Rail Roads must sooner or later be the only safe and permanent communication from our western waters to the Mississippi."[67]

The railroads would come, but not for decades. In the meantime, legislators pressured by the worsening navigational problem at Old River needed a reachable remedy. When Hébert asked for an appropriation to keep dredging at the mouth of the Red, he told the legislature that dredging alone would not keep Old River open. He advocated damming the Atchafalaya outflow to turn the Red back toward the Mississippi. The state engineer believed that this measure, which he knew was politically untenable, was preferable to making the Raccourci cutoff.[68]

Exasperated by the problem and unwilling to consider closing the Atchafalaya, the Louisiana legislature approved $6,000 for the Raccourci cutoff in March 1847, giving the state engineer six months to complete the work. If the amount appropriated proved inadequate, the legislation "authorized and empowered" private citizens to finish the job "at his or their own expense."[69] By midsummer, the heat at the worksite withered Hébert's small crew of twenty-three state hands. In the end, a private contractor, a Mr. D. Hoard, stepped in to complete the cut by the close of the year. The Raccourci trench, fifteen feet deep and twenty feet wide at the bottom, with a width of thirty feet across the top, began slowly capturing the Mississippi's current in early 1848.[70]

♦ ♦ ♦

"IT WAS A GRISLY, HIDEOUS NIGHT, and all shapes were vague and distorted." So begins Mark Twain's relation of a story about the Raccourci cutoff. Twain was piloting riverboats on the Mississippi in the 1850s, and the south end of the horseshoe lake tracing the old Raccourci river bend was still accessible to vessels passing up and down the river. As the story goes, a pilot navigating in the darkness accidently steered his steamboat into the old channel, where he became lost among the sandbars and shoals. Twain concludes the tale in classic ghost yarn fashion:

> So to this day that phantom steamer is still butting around in that deserted river, trying to find her way out. More than one grave watchman has sworn to me that on drizzling, dismal nights, he has glanced fearfully down that forgotten river as he passed the head of [Raccourci] island, and seen the faint glow of the specter steamer's lights drifting through the distant gloom, and heard the muffled cough of her 'scape pipes and the plaintive cry of her leadsmen.[71]

Not to spoil Twain's story, but Raccourci Bend wasn't "forgotten" that quickly. As Hébert predicted, the redirected Mississippi provided relief to around thirty plantations with waterfront property around Raccourci Bend. (Like the cutoff at Turnbull Bend, Raccourci Bend became Raccourci Island.) These were the planters who had stood by helplessly seventeen years earlier as the Mississippi roared through Shreve's cutoff and carried away part of their cotton land. After the Raccourci cutoff began to capture the Mississippi, these plantations enjoyed still water at their landings and, for several years at least, convenient access to the main channel. Sedimentation studies of Raccourci Old River indicate that silt began filling the lake soon after the cutoff. The upstream end of the horseshoe probably closed first, while the downstream end may have remained navigable, at least during times of high water, until the 1880s.[72]

To the relief of the planters over on Tunica Bend, the Mississippi did not accept the Raccourci channel rapidly. Unlike Shreve's cutoff, which captured the entire river in a matter days, the river didn't fully invest in the Raccourci cutoff for several years. Factors slowing the diversion included the substantial clay deposits that Dunbar noted, the length of the trench (close to a mile), and its relatively shallow depth; Shreve's cut was deeper by seven feet. By the early 1850s, however, the Mississippi's main force was flowing down the cutoff, falling a vertical distance of two to three feet in one mile instead of nineteen. Delayed but not deflected, those "most mischievous consequences" predicted by the legislative panel of engineers back in 1846 came lapping at the levees along Tunica Bend and south toward Pointe Coupée.[73]

Although the 1844 resolution authorizing the Raccourci cutoff included a process for compensating landowners through whose property the channel would cross, the bill said nothing about reparations for downstream damage. Echoing the sentiments of many, a *Times-Picayune* editorialist urged the legislature to provide affected property owners below the cutoff with "a reasonable indemnity." The state's response was not swift. While the Mississippi gouged and hammered its new channel, Louisiana's seat of government moved from New Orleans to the new Gothic-towered statehouse in Baton Rouge. Once ensconced in their new chambers, lawmakers haggled for years over bills to settle with ruined landowners, accomplishing little of substance until 1858. That year, the state agreed to pay one of the many downstream planters claiming damages, setting a precedent for other claims. But as aggrieved planters began to realize some relief, a greater conflict swept away their fight with the Louisiana legislature.[74]

◆ ◆ ◆

THE CIVIL WAR'S CALL TO ARMS coincided with Louisiana's victory after a thirty-year battle with the Atchafalaya raft. Shipping interests in the Atchafalaya Basin cheered the long-awaited connection to the Mississippi and the Red, even if the route through the Old River confluence was only safely navigable during the high water season from December through May. On the other side of the equation, the raft's removal further increased the Atchafalaya's carrying capacity, hastening the Red River's divorce from the Mississippi. The Mississippi passing into the Atchafalaya flowed through Old River's lower arm, maintaining a tenuous connection with the Red River. More evident, and disappointing to many, was the continued accumulation of sediment in the upper arm of Old River.[75]

As the Atchafalaya captured more of the Red River's flow, Old River achieved a curious state of equilibrium. Engineers and local river folk noticed that the water in Old River's lower arm sometimes flowed from the Mississippi toward the Red and Atchafalaya, while at other times the flow reversed, bringing part of the Red into the Mississippi. Much like the bubble in a carpenter's level, the water in Old River responded to the push and pull of the three rivers. During times when water on the Red rose higher than the plane of the Mississippi, the Red's flow ran eastward through Old River. Once the high water on the Red began to subside below the level of the Mississippi, the balance between the three rivers tipped westward again, sending the flow in the opposite direction.[76]

Taking turns forcing water through Old River's lower arm, the Red and Mississippi each contributed sediment to the narrow channel. Observations indicated that Old River's current moved more swiftly when flowing from west to east, sometimes with enough force to clear away some of the shoaling. Unfortunately, the buildup of sediment outpaced any natural scouring effect. Navigation through the critical junction, even during high water periods, depended upon continuous dredging. Nature seemed intent upon weaning the Mississippi away from its western tributary and reconnecting the Red with its former course down through the Atchafalaya Basin.[77]

Among the people down the Atchafalaya River and along bayous Teche, Lafourche, and Plaquemine, over in New Orleans, up the valleys of the Red and Ouachita, and in the halls of the Louisiana statehouse in Baton Rouge, the rift between the rivers riled two opposing factions, one urging state and federal authorities to keep Old River open at all cost and the other pressing those same officials to side with nature and close it. The main issues were flood control, navigation between rivers, and the conversion of swampland into fields of cotton and sugarcane. Ultimately, the decision would fall to a new authority of combined civilian and military expertise called the Mississippi River Commission.[78]

CHAPTER 3

Blessing and Curse

"A blessing of incalculable advantage" is how an anonymous *Times-Picayune* editorialist in 1877 characterized the economic benefit New Orleans derived from the confluence of rivers at Old River.[1] The writer was referring to river-borne commerce with the valleys of the Atchafalaya, Red, and Ouachita made possible by the Old River connector; however, he might just as well have been talking about another, even more urgent, blessing. Old River's natural outlet to the Atchafalaya River siphoned away a significant percentage of the Mississippi's flow; by 1910 that percentage was up to 18 percent and would continue to climb. Old River's potential to pull that much water out of a flood was, indeed, a blessing for the Crescent City.[2]

Between the turn of the nineteenth century and the 1870s, the Lower Mississippi Valley experienced at least sixteen major floods. The Mississippi River invaded the streets of New Orleans during eight of those episodes, with the most destructive rises coming in 1816 and 1849. Historical geographer Richard Campanella wrote that the 1849 deluge "ranks as New Orleans' worst flood until the Hurricane Katrina levee failures of 2005." Some of these inundations came from collapsed dikes adjacent to the city, while others resulted from Mississippi floodwaters spilling into Lake Pontchartrain and advancing on New Orleans from the rear. So with good reason, the city's residents watched each spring rise with trepidation. Most of the time, the levees along the city's long waterfront from the lower suburbs (present day Lower Ninth Ward) up to

Carrollton held, but poorly constructed levees protecting plantations upriver often failed. In the catastrophic flood of 1858, privately maintained levees throughout the valley upstream from New
Orleans crevassed in forty-five places, covering fields and farming
communities. These crumbling levees, augmented by the Atchafalaya distributary, helped to move water out of the Mississippi's main
channel and keep the Crescent City dry.[3]

As the nineteenth century entered its final quarter, however, a
seminal change in Lower Mississippi Valley flood control would
eventually make the levees above New Orleans more secure and
make the Old River outlet even more vital to downstream interests.
The Civil War wrecked the southern river states' ability to maintain
their protective earthworks, and in 1874 a reunited Union looked
on in sympathy as yet another devastating flood laid waste to much
of the meager postwar repairs. For decades, the US government
had stubbornly limited its involvement with the Mississippi and its
tributaries to navigational improvements (Shreve's snag-pulling for
example), because constitutional arguments forbade using public
funds to protect the interests of riverside towns and plantations.
Following the 1874 inundation, that policy began to shift. A pres-
identially appointed commission of army and civilian engineers,
the precursor to the Mississippi River Commission, convened to
investigate ways to bridle the Mississippi River. As it turned out,
commission members (one of whom was former state engineer
and Louisiana governor Paul O. Hébert) already had a blueprint for
their task.[4]

Following the disastrous 1849 flood and desperate calls for
federal action from affected states, Congress funded two extensive studies of the Lower Mississippi River, one by the US Corps
of Topographical Engineers, headed by Captain Andrew A. Humphreys and Lieutenant Henry L. Abbott, and the other by a private
civil engineer, Charles Ellet Jr. Ellet's report, issued first, is essen-
tially the same approach to Mississippi River flood control adopted
by the Corps of Engineers in the twentieth century: sturdier levees,

improved natural outlets to disperse the floodwaters, and reservoirs to restrain some of the Mississippi's tributaries. The report by Humphreys and Abbott considered several alternatives, including digging new cutoffs, diverting tributaries and constructing reservoirs, opening artificial outlets, and building bigger levees. With a disdain for Ellet's conclusions bordering on ridicule, the two topographical engineers found fault with every approach to flood control save one. After a rambling, name-dropping discourse about European rivers such as the Rhine, Po, and Arno, the authors concluded confidentially that only "levees . . . may be relied upon for protecting all the alluvial bottom lands liable to inundation below Cape Girardeau."[5]

Compared to Ellet's diversified approach, Humphreys and Abbott offered nervous lawmakers and their constituencies a simple solution that, disastrously, became the framework for major flood control decisions reaching into the twentieth century. While not everyone was convinced, the "levees-only" mantra quickly found its way into state and federal policies and procedures. One group that may have viewed the sole reliance on levees with some concern, especially given the authors' opposition to outlets, was the residents of New Orleans. Outlets, the two army engineers contended, caused sedimentation, raising the riverbed downstream by reducing the Mississippi's volume and force. To the relief of riverside communities below Old River, however, Humphreys and Abbott made strong exception to the Atchafalaya distributary. Their report declared the distributary to be an "efficient" natural outlet and warned of "disastrous consequences" should Old River be closed.[6]

The argument for closing outlets certainly resonated with the people of the lower Atchafalaya Basin, who suffered when Mississippi floodwater inundated the Grand River and Grand Lake region via Plaquemine Bayou. To the south, Bayou Lafourche also carried too much water when the Mississippi was in flood, wreaking havoc on the levees down in the Lockport area. Joining forces with these flood-weary residents of south Louisiana was a relatively new and powerful constituency: the railroads. By the end of the 1860s, tracks spanned the United States from east to west, and eager railroad corporations pushed into ever more remote places to exploit the demand for freight and passenger service. In the Atchafalaya Basin, closure of the Mississippi outlets at bayous Plaquemine and Lafourche offered a good opportunity for railroad lines to displace the Teche to New Orleans water-borne shipping business.[7]

Still another case against outlets, including the one at Old River, came from James Buchanan Eads, a self-taught engineer who bridged the Mississippi at St. Louis and famously deepened the channel at the Mississippi's mouth in 1876, removing a long-standing hazard for shipping. Eads's success in increasing the river's depth at the mouth's South Pass from nine to twenty feet came from the use of jetties, piers composed of vertical pilings and baffles of interwoven pine and willow branches extending from the bank at a downstream angle out into the current. In theory, the jetties concentrated the river's force at midstream to abrade and lower the riverbed. Eads contended that outlets, by lessening the river's volume, decreased the jetties' effectiveness. Humphreys, among others, denounced the scouring capability of both jetties and levees, citing the presence of rock-hard clay throughout much of the Mississippi's bed. Although Congress and the Corps of Engineers rejected Eads's plan to employ jetties throughout the Lower Mississippi, the notion of manipulating the river channel so that it entrenched itself persisted, later becoming erroneously attached to the levees-only doctrine.[8]

The simplicity and political neutrality of relying solely on levees became a torch picked up and held high by the Mississippi River Commission, authorized by Congress in 1879 to take over the great river's navigational improvement. Tiptoeing around the nagging constitutional questions about spending federal dollars to protect private property, the commission included levee-building in its scope of work from the beginning. The new commission mirrored the 1874 presidential commission in its mix of army and civilian appointees, satisfying those who felt that the Mississippi should not be entirely under military control. In the working relationship that continues today, planning falls to the Mississippi River Commission, while the Corps of Engineers handles most of the research, construction, and maintenance. If Humphreys and Abbot provided the commission with an easy solution to flood control, needing only money, men, mules, and dirt, their levees-only report offered no advice in solving the dilemma at Old River beyond the authors' admonition to keep the three rivers link open.[9]

♦ ♦ ♦

THE MISSISSIPPI WAS RAISING A FLOOD at the end of December 1881. After a brief fall during December 6–27, the river swelled steadily over the next three months into a record-breaking overflow, finally cresting on March 27, 1882. The Mississippi River Commission's assistant engineer, John Ewens, proclaimed the 1882 flood "the greatest flood discharge ever measured on the Mississippi River." The young engineer could speak with authority; he witnessed and documented the Mississippi's prodigious rise that year from a front-row seat on the banks at Old River. Assigned by the commission to conduct a thorough study of the three rivers area, Ewens and his small party of technicians descended the Mississippi just ahead of the flood. Leaving St. Louis in late November, the group arrived at the town of Red River Landing on December 3, 1881.[10]

In those days, Red River Landing was a small community boasting a store, hotel, and church, along with several houses. New York artist Julian Oliver Davidson visited the Old River area the year after Ewens was there and produced a lively composition of post-card scenes of the town and surrounding landmarks for *Harper's Weekly* (see Figure 3.1). Located on the Mississippi River about three-fourths of a mile south of the mouth of the lower arm of Old River, Red River Landing was a busy steamboat stop for vessels plying the Mississippi and the Red. When Ewens was there, the town also had a post office. With its close proximity to the Angola Ferry,

Figure 3.1 Old River area scenes in 1883 depicted in J. O. Davidson's illustration for *Harpers Weekly*, April 14, 1883. Courtesy of the Historic New Orleans Collection, 1974.25.30.724.

the landing linked river travelers to roads reaching out west and east to intersect with the railway lines spidering across the maps of Louisiana and southwest Mississippi. Today, Red River Landing is a key Mississippi River gaging station, but all traces of the frontier hamlet have been lost.[11]

Ewens and his assistants spent twelve months at Old River conducting the first systematic investigation of the three rivers confluence. In addition to their detailed study of the lower arm connecting the rivers, the crew examined the Mississippi, Atchafalaya, and Red Rivers along with nearby Bayou des Glaises. The previous year, Major G. W. Howell had directed the Corps of Engineers' first survey of the Atchafalaya River from Berwick Bay to Old River. His maps and Ewens's Old River data formed the baseline engineering study of the three rivers confluence.[12]

Using two steam-powered vessels, a launch, and a catamaran, in tandem with onshore transit stations, Ewens meticulously gathered data on discharge volumes, stream velocities, water temperature, sediment analyses, and weather conditions. Ewens noted some of the hazards his crew faced each day out on the water, including "the continued presence of passing boats" and drifting trees. Despite the natural and man-made interference, the crew took soundings at regular intervals throughout the year to establish the fluctuating depths and changing cross-section profiles of these streams. In the Mississippi, the soundings were particularly difficult due to the tangled mass of submerged trees and limbs continuously tumbling along the river bottom. On days when inclement weather interrupted the routine, the crew stayed busy inside the expedition's on-site "shop" making adjustments and repairs to the scientific equipment they used.[13]

Ewens reported that when the Mississippi carried the "great flood" (Ewens's words) of '82, the Red remained at near normal levels. Work continued despite the overflow. As the Mississippi rose, transit operators on shore establishing the daily transects for soundings and other measurements had to operate from platforms

in trees. As usual in times of flooding, floating debris filled the
Mississippi, doubling the danger for the men on the catamaran
and launch. Looking back, and considering the flood along with
the daily hardships at times of normal flows, Ewens laconically
described his year at Old River as "most trying" and proudly noted
in his report that his team had suffered no injuries.[14]

During the flood, the engineers kept watch on a potential trouble spot a short distance north of Old River. Here, where the Mississippi makes a westward loop, sweeping around a sharp curve
known as Widow Graham Bend, where only about five miles of
vulnerable bottomland separate the Mississippi from the Red River.
Since the 1850s, the Mississippi had sometimes overflowed its west
bank at Widow Graham Bend at a spot called the Bougere Crevasse.[15] Just below the bend, a sluggish little stream known as Grand
Cutoff Bayou traced a former connection across the narrow stretch
between the two rivers. To Ewens's relief, the Mississippi's floodwaters here weren't threatening to cross over into the Red River, at
least not this time. In the book's last chapter, we will revisit Widow
Graham Bend and examine the Bougere Crevasse and Grand Cutoff Bayou in more detail.[16]

Ewens provided the Mississippi River Commission with hard
data about the hydrology and sedimentation at Old River. His report
includes a concise account of the interaction between the periodic
reversal of flow direction, mentioned earlier, and the dynamic role
played by Turnbull Island. Although riverboat pilots and local residents were familiar with the changing flow direction through Old
River, Ewens's study gives a close look at the phenomenon:

> The first sign [of a direction change] manifest is a marked check-
> ing of the current in Old River, and a more marked degree of
> clearness in the water. These changes are very easily noticed on
> the first day. On the second day the current nearly ceases. On the
> third [day], the change in flow is found to have taken place; the
> current will be found as swift as usual, and the water will have the
> usual mud color.[17]

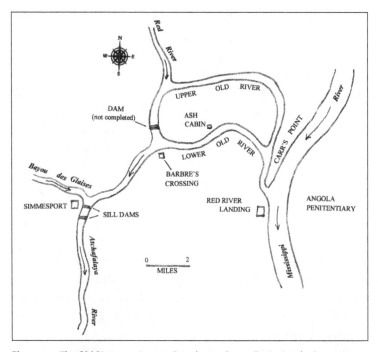

Figure 3.2 The Old River area in 1882. Based upon Reuss, *Designing the Bayous*, 82; "Trip on the Atchafalaya River, Drawn by J. O. Davidson." *Harper's Weekly*, April 14, 1883, 237.

A critical effect of the flow reversal, revealed by the change in the water from muddy to clear, is the deposition of sediment when the direction change occurs. The "clearness in the water" indicates that the slowing (checking) current was losing its sediment load, and the return to "the usual mud color" was a sign of accelerated flow with increased sediment transport. After all, sedimentation was the Mississippi River Commission's chief concern at Old River; by the late 1880s, thousands of federal dollars were being spent each year on dredging to keep the pass open. In Ewens's opinion, it was the back-and-forth movement of water around Turnbull Island, the former river bend severed by Shreve, that facilitated the shoaling: "[The island] is in fact the nucleus of all the sand bars in the vicinity."[18]

Daily observations of Old River's lower arm over the course of a year allowed Ewens to draw some important conclusions. Sediment was accumulating in three places along this reach: Barbre's Crossing, at Old River's junction with the Red; Carr's Point, where Old River joined the Mississippi; and Ash Cabin Bend, a dogleg reach about midway through the passage named for a homestead seated on the north bank of Old River's lower arm (see Figure 3.2). The team's daily measurements revealed just how the direction of flow through the channel affected the shoaling at these spots. Compared to the Mississippi, the Red brought a relatively small amount of sediment into Old River when the current flowed from west to east. Ewens noted that the Red River's reddish color came from fine-grained particles, which he characterized as being "held in solution" as opposed to larger particles "held in suspension." The light sediment load minimized shoaling and the swifter current when the flow was toward the east helped the dredging operation with some modest scouring, but Ewens found that the eastward flow had a negative effect, as well.[19]

Carrying a heavy sediment load into Old River, the Mississippi hooked around Carr's Point at an acute angle and lost much of the momentum from the main current. Further slowing around Ash Cabin Bend caused the shoaling there, and the creeping current deposited much of its remaining sediment at the junction with the Red River. But as troublesome for navigation as these two spots were, the shoaling at the mouth of Old River, where it encountered the Mississippi, worried Ewens the most. This problem was caused by the west to east flow, the direction associated with a light sediment load and beneficial scouring action. When the flow from the Red left Old River and merged with the Mississippi adjacent to Carr's Point, the incoming water caused enough drag on the Mississippi to cause eddying and shoaling. As a result, a sandbar at the snout of Carr's Point was steadily growing southward. Ewens's team observed that the bar lengthened three hundred feet during 1882.

Should the trend continue, Ewens feared, the Carr's Point extension could permanently close Old River.[20]

◆ ◆ ◆

A CURIOUS ARMADA gathered in August 1887 along the wharf at Simmesport, Louisiana, a town about four miles down the Atchafalaya from Old River (see Figure 3.2). Its steam-whistle tootling, the Corps of Engineers' sternwheeler *General Newton* served as flagship for a convoy of two steam launches, seventeen barges, and five quarter boats, which would serve as floating dormitories for a small army of laborers. The government invasion had come to construct a sill dam, a submerged structure spanning the channel bottom. The Atchafalaya River's increasing threat to capture the Red River's full flow warranted action. If the Mississippi River Commission's strategy proved successful, this sill dam and five more like it placed at intervals down the river would stop the deepening of the Atchafalaya's bed, while permitting the river's normal flow and boat traffic. In a clearing about five hundred feet below the mouth of Bayou des Glaises, workers set about drilling an artesian well for the crew's drinking water and building a double-walled "ice house," insulated with sawdust, for storage of perishables. Town merchants and entrepreneurs watched all this activity with anticipation; the arrival of the Corps of Engineers promised federal greenbacks aplenty for the next nine months.[21]

Local whiskey peddlers fared especially well. W. G. Price, the engineer in charge of the work, acknowledged that eight of his men died after drinking bad liquor and sleeping on the ground. For those laborers who managed to show up for work each day, the job was arduous. Built in much the same way as Eads's jetties, the base of the dam comprised overlapping "mattresses" made of willow brush sandwiched between willow poles, all enclosed in a pine frame. Each mattress was three feet thick and measured 304 feet by

75 feet. The men heaved these unwieldy contraptions off the sides of the barges and piled 450 tons of rock on each one. Atop the sill, the dam proper was built up of alternating layers of mattresses and gravel mixed with clay. A dip in the center of the span allowed room for boats to pass over without dragging.[22]

The sill dams on the Atchafalaya were part of an elaborate strategy to assert control over the three rivers confluence. Aside from dredging, nothing man-made had been done at Old River in the half century since the Shreve and Raccourci cutoffs. All concerned generally agreed that the steady enlargement of the Atchafalaya's carrying capacity was a leading factor in the progressive separation of the Red and Mississippi Rivers. At the other end of Old River, the Corps of Engineers watched the ever-changing sandbars in the Mississippi's channel. Even more menacing than the Carr's Point bar was a steadily growing sandbar on the east side of the Mississippi, emerging in front of the broad floodplain occupied by the newly established Angola penitentiary. If it continued to expand, the Mississippi River Commission reported, the Angola bar might crowd the Mississippi to the point of diverting the great river westward through Old River and down the Atchafalaya. In its 1888 report to the secretary of war, the commission noted the disastrous consequences for New Orleans should this occur. (Crescent City residents likely found little solace in the report's conclusion that, should the channel jump occur, Atchafalaya Bay would be a poor substitute for New Orleans.) The report points out that Old River, having once been the main channel of the Mississippi, was of sufficient dimensions to carry the river once again. The same couldn't be said for the Atchafalaya yet, so the proposed sill dams took on added significance in the Angola bar scenario.[23]

While the Mississippi River Commission wrestled with the evolving channels at the three rivers confluence, the crossroads remained a busy commercial waterway. Much depended upon the ability of steamboats and other traffic to navigate the Old River passage. Over the course of a typical year, cotton, coal, livestock, farm produce and dairy products, grains, scrap iron, bricks, and sawmill-bound rafts of valuable timber were but a few of the commodities moving through the seven-mile corridor.[24] In times of high water, both the upper and lower arms were passable. As the water level fell, the upper arm closed and only continuous dredging kept Old River's lower arm open. To augment the dredging, the Corps employed a technique known as washing, using the *General Newton's* churning paddlewheel to stir up the riverbed sediment so the current could carry it away. At the worst trouble spots, the Corps moored the *General Newton* to guide piles, heavy posts driven into the Old River bed, so the paddlewheel could run in place over a particular shoal area. Another strategy, ultimately unsuccessful, was the use of spurdikes, or temporary jetties, to speed the current enough to counteract the shoaling.[25]

When the water level fell, the shoals at Ash Cabin Bend tested the riverboat pilots' patience and skill. Here, the riverbed was hard clay instead of sand and silt, making dredging difficult and passage tricky in low water. In an efficient use of government resources, the Corps mined clay here for the fill in the sill dam construction. Yet despite the engineers' efforts, dropping river levels sometimes closed the lower arm for several weeks at a time. In November 1888, with near-record low water on both the Mississippi and the Red, Old River dried up completely. A Mississippi River Commission engineer reported that "[wagon] teams were crossing it on a dry, dusty road."[26]

◆ ◆ ◆

THE SCHEME that included the Atchafalaya sill dams was ambitious. To manage Old River, the Mississippi River Commission decided to cobble back together the three rivers connection to a sort of pre-Shreve's cutoff configuration. In the plan designed for what the commission called the "rectification" of the Mississippi and Atchafalaya Rivers, the Corps of Engineers would have to juggle three balls in the air to (1) stop the separation of the Red and Mississippi, (2) maintain a navigational connection through Old River, and (3) allow part of the Mississippi to continue to drain into the Atchafalaya River. A complicated mission, to be sure. The commission had to navigate tricky political as well as hydrological shoals. From the top of the Tensas Basin down to the New Orleans riverfront and from Bayou Teche across the Atchafalaya Basin to Baton Rouge, merchants; planters; bankers; brokers; shipping and railroad concerns; and, of course, politicians, had a keen interest in what was happening at Old River.[27]

Before the Mississippi River Commission rolled out its blueprint for setting things right, the commission considered and rejected several proposals. The most radical plan came from former commission member James B. Eads. Still hewing to his controversial theory that the Mississippi would entrench itself if its force was not weakened by outlets, Eads advised closing the Atchafalaya distributary with a high dam to reestablish the Red River as a Mississippi tributary. He insisted that the combined power of the two rivers would scour and deepen the Mississippi's channel. Although the immediate result would be higher water, requiring taller levees, Eads believed that the scouring would eventually lower the river's height downstream. Objections were many. Besides severing the commerce with the Atchafalaya and its connecting streams, the closure of the Atchafalaya outlet would send too much water down to New Orleans, especially in times of high river levels on both the Red and Mississippi. Moreover, the high dam would be expensive and, should it break, the Atchafalaya Basin would be in grave danger.[28]

A plan given more consideration came from Major Amos Stickney, the district engineer in the Corps of Engineers' office in New Orleans. He proposed closing Old River's lower arm and dividing the Red with a low water dam at the head of Turnbull Island, sending part of the flow down the Atchafalaya and the rest through the upper arm to the Mississippi. Another idea, eventually discarded, was to close Old River and construct a lock or canal connecting

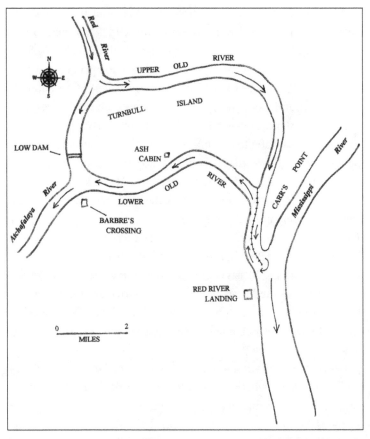

Figure 3.3 The Mississippi River Commission's 1889 plan for stabilization of the Mississippi, Red, and Atchafalaya Rivers' confluence. Based upon Annual Report of the Chief of Engineers, 1889, Part 4, 2724–41, Plate No. 3 Plan and Section of a Jetty for Improving the Mouth of the Red River.

the Red and Mississippi. One of the locations scouted for this cut was the Grand Cutoff Bayou and Bougere Crevasse weak spot a few miles above Old River that had worried Ewens during the flood of 1882. While the canal and lock plans had some merit, they both called for closing the Atchafalaya outlet, which New Orleans interests strongly opposed.[29]

With much of the Lower Mississippi Valley at the bargaining table, the Mississippi River Commission's Old River plan was a grand compromise. Along with the series of sill dams below Simmesport to stop the Atchafalaya's enlargement, the plan endeavored to turn Old River into the hydrological equivalent of an automobile traffic roundabout (see Figure 3.3). A low dam on the west side of Turnbull Island, between the Red and Atchafalaya, would allow some Red River water to continue down the Atchafalaya while sending part of the flow—that part blocked by the dam—down to the Mississippi through Old River's upper arm. At the toe of Turnbull Island, where the upper and lower arms converged, a jetty would extend about eight thousand feet down the middle of the Old River channel and jut out into the Mississippi's current. The jetty would divide the channel in half, allowing the water from the Red River to stream out into the Mississippi without forming a shoal. According to the design, as the Mississippi's current swirled around the end of the jetty, gravity would draw a portion of the river into the other side of the Old River channel, to be conveyed to the lower arm and on to the Atchafalaya.[30]

The commission's design for Old River seemed to meet the needs of all concerned; whether it would have worked remains a mystery. The Corps of Engineers completed only two of the proposed sill dams, while the dam west of Turnbull Island was left unfinished due to funding reductions brought on by the Panic of 1893 and the Spanish-American War. The 1.5-mile-long jetty at the toe of the island never left the drafting table. During the first years of the twentieth century, regular dredging kept the Old River passage open most of the time. To the riverboat pilots, the three rivers confluence seemed to have stabilized; however, the political turbulence surrounding Old River doggedly hung on to the Mississippi River Commission's agenda.[31]

◆ ◆ ◆

WHILE THE MISSISSIPPI RIVER COMMISSION sought a way to prevent natural forces from closing Old River, work began on a Mississippi River lock at the head of Bayou Plaquemine, about ninety river miles to the south. Like the Atchafalaya distributary at Old River, Bayou Plaquemine and Bayou Lafourche further downriver carried Mississippi River water into the Atchafalaya Basin, serving as avenues of commerce as well as openings for floodwater intrusion. When the removal of the Atchafalaya River raft gave the lower Atchafalaya Basin a viable alternate route for commerce through Old River, the state of Louisiana temporarily dammed Bayou Plaquemine. By that time, railroads in the Atchafalaya Basin made it possible for many planters to ship via rail to New Orleans. However, the loss of the bayou's water in the interior basin hindered the use of the interconnecting system of waterways. In 1882, Congress authorized the construction of the Plaquemine Lock. The Corps of Engineers dammed Bayou Lafourche in 1904 when the Plaquemine Lock was nearing completion.[32]

When the Plaquemine Lock finally opened in 1909, citizens in the Atchafalaya Basin and up the valleys of the Red and Ouachita seized the opportunity for a long-awaited separation of the Red and Mississippi Rivers at Old River. After all, the new lock provided a reliable shipping route between the Mississippi and the rivers to the west, making Old River's navigational function unnecessary. And big floods on the Mississippi in 1903 and 1908 were reminders of the trade-off that Old River had always demanded. The passageway for commerce gave the great river a channel-wide gap through which the deluge poured. Now, for the first time in history, the way was clear to shut the door at Old River and stop the Mississippi from troubling its western neighbors.[33]

Divorcing the rivers dominated the agenda at a public meeting in Tallulah, Louisiana, held on March 11, 1909, although those attending expected more than flood protection from the separation of the Red and Mississippi. The gathering represented Louisiana's

Fifth Levee District, the northeastern parishes of East Carroll, Madison, Tensas, and Concordia, a swath of rich delta land stretching up through the Tensas Basin all the way from Old River north to the Arkansas state line. If the Mississippi River Commission consented to closing Old River once and for all, planters who owned land along the Tensas River and its interconnected bayous could drain and cultivate over a million acres of swampland. The key to realizing this prosperity was lowering the head or elevation of the river surface into which their property drained.

In the district of these planters, sometimes called the Red River Backwater, rainwater runoff drained west and south, down the Tensas to Jonesville, where the Little, Tensas, and Ouachita Rivers join to become the Black River, and from there on southward down the Black to flow into the Red River just above Old River. Recalling the reversing current phenomenon at Old River, when the surface of the Red was higher than that of the Mississippi, the Red's high water mark at Barbre's Crossing at the head of the Atchafalaya River never gaged over forty-eight feet above mean Gulf level. In contrast, higher flows on the Mississippi brought enough water through Old River to cause the Red to back up, sometimes reaching the fifty-five-foot mark on the Barbre's Crossing gage.[34]

Even though some of their properties lay over one hundred miles north of Old River, the planters of the Fifth Levee District could envision a profitable expansion of their cultivatable acreage after the separation of the rivers. In the Mississippi River floodplain, where variations of just a few inches in topography mean the difference between wet ground and dry, a lower head of seven feet would transform the landscape; much of the property normally under water during Mississippi River floods would be spared the overflow. Since 1882, a span of time well within the collective memory of the people at the Tallulah meeting, the big Mississippi floods came on average every three or four years.[35] During the major Mississippi River flood of 1912, the Mississippi River Commission inspected the

Tensas Basin area and confirmed the levee district's estimate of the land to be reclaimed after the separation.[36]

To the south in the Atchafalaya Basin, the Atchafalaya Protective Association also coalesced around the separation issue. Ten parishes in south Louisiana saw the Red-Mississippi divorce as a long-awaited salvation from over a century of suffering every time the Mississippi's floodwaters came surging west through Old River. The group's web of alliances included several Louisiana Levee Districts, members of Congress, and powerful railroad corporations, whose tracks traversed the Atchafalaya Basin linking New Orleans with Dallas and Houston.[37]

In making its appeal for protection, the Atchafalaya group claimed a higher purpose for the separation by aligning itself with Eads's theory about closing outlets to allow the river to entrench itself, sometimes referred to as the Confinement Theory.[38] By the 1880s, the Corps of Engineers had adopted this doctrine, declaring with confidence that "[the Mississippi River] flows through a bed adjustable by its own forces to its own needs."[39] Closing Old River, argued the Atchafalaya folks, would simply confine the river according to the accepted model. To further strengthen its argument, the group felt compelled to remind the commission about the danger of the Mississippi changing course and moving into the Atchafalaya Basin: "The calamity which would follow such diversion of the Mississippi would not be confined to the fact that New Orleans would become an inland city, but the most beautiful and fertile valleys of our beloved State would be devastated by flood."[40]

In this instance, the separation proponents' hyperbole was unnecessary. The Atchafalaya Protective Association wasn't telling the Mississippi River Commission anything its members didn't already know.[41]

◆ ◆ ◆

THE MISSISSIPPI RIVER RAISED RECORD-SETTING FLOODS
in 1912 and 1913,[42] while the divorcement issue remained unresolved
on the commission's meeting table. South of Memphis, it seems
everyone held an opinion on the subject of Old River. At river
landings like Vicksburg, Natchez, and New Orleans, the debate
dominated public meetings held aboard the Mississippi River
Commission's steamer *Mississippi* during the commission's yearly
low-water and high-water inspection voyages.

Pro-separation resolutions came from Louisiana's bayou country west of the Mississippi, from town councils, parish police juries
(Louisiana's equivalent to county boards of supervisors), and levee
districts calling for Old River's closure, shutting out the Mississippi
and reuniting the Red and Atchafalaya rivers. These communications frequently cited the opening of the Plaquemine Lock, heavy
local taxes paid for levees that Mississippi River floods repeatedly
demolished, and estimates ranging from thousands to millions of
acres that could be reclaimed after Old River is dammed. A promise
of "large investments in capital" arrived from the Louisiana Timber Corporation (with an office on Wall Street in New York City),
banking on land to be reclaimed from frequent overflows after the
divorcement. The Texas and Pacific Railroad, speaking in favor of
the separation, touted the general benefits to be enjoyed by the
people through land reclamation. Quietly, the railroad also stood
to profit from the disruption of river commerce that the Old River
closure would cause, inducing planters and merchants to switch
their shipments from boats to rail cars.[43]

On the other side of the issue, antiseparation proponents made
their case for maintaining the status quo. The Natchez Chamber
of Commerce, New Orleans Board of Trade, Baton Rouge Board
of Trade, and Shreveport Chamber of Commerce objected to
the radical change in the well-established pattern of river traffic
through Old River. They feared what might happen to commerce
if the Plaquemine Lock (or a new lock at Old River) should fail.
Also, locks create delays when boats have to wait their turn to pass

from one river to another. Old River, with its occasional periods of low water, certainly wasn't ideal, but riverboat pilots knew when to expect that kind of problem. The Steamboat Traffic Association argued that closing Old River and redirecting shipping through the Plaquemine Lock would increase the number of hazardous bridges the boats would have to negotiate ascending the Atchafalaya River to reach the Red and Ouachita systems. The heightened possibility of damaging boats and bridges would ratchet up insurance costs. Boat owners stressed that this additional expense, plus the disruption of longtime shipping routes, would give an unfair advantage to the railroad companies.[44]

One railroad firm that opposed closing Old River was the Louisiana Railway and Navigation Company, citing the expense of having to move track and construct new bridges. Of course, the levee districts and parishes on the Mississippi below Old River feared what future floods would be like without the Atchafalaya outlet to deflect the blow. New Orleans interests pressed the Mississippi River Commission, Congress, and the Louisiana legislature to keep the outlet open. On July 5, 1910, Louisiana governor J. Y. Sanders signed Act No. 113, opposing the separation of the rivers, stating simply that the loss of the Atchafalaya outlet would "threaten the safety of the country below Red River."[45]

◆ ◆ ◆

WHILE THE SEPARATION DEBATE ROILED, the Mississippi River Commission dispatched Assistant Engineer E. J. Thomas to Old River to determine the best site for the dam should the decision fall that way. Thomas also surveyed the area just north of Old River for a lock and canal site to maintain a navigational connection through the area. Updating Ewens's eighteen-year-old study and more recent cross sections mapped in 1895 and 1904, Thomas documented the ever-changing profiles of the riverbeds and banks. His findings included a report that the Atchafalaya River was still increasing its depth and width, perhaps slowed somewhat by the sill dams. The upper sill dam, closest to Simmesport, even registered a slight decrease in depth since it was last profiled in 1904. Over on the Mississippi, the Carr's Point sandbar now extended 2.5 miles and the widening bar across the river at Angola had constricted the channel by as much as two-thousand feet since the last soundings.[46]

To effectively close Old River, the commission needed a dam eighty-five feet high, with a base three hundred feet wide, positioned somewhere along the two-mile-long chute from the toe of Turnbull Island out to the Mississippi. For additional buttressing, Thomas called for filling in the Old River channel on both sides of the dam for a distance of "several hundred feet." The structure's stability depended upon what lay below the chute's bed. Accordingly, Thomas and his crew used a pile driver to collect twenty core samples along the chute at depths of up to 113 feet below mean Gulf level. All of the coring columns registered the presence of thick sand deposits interspersed with layers of clay and gravel, not unexpected given the presence here of superimposed beds representing old Mississippi River channels. Despite finding sand in 75 percent of the core samples, Thomas remarked that the area contained "firmer substrata than expected." The estimated cost for the dam came in at $407,790.[47]

For the canal and lock, Thomas suggested two possible locations. One was about five miles upriver, just below the landing for the Black Hawk plantation, and the other a little over four miles further north at a spot called Union Point Landing. The Union Point canal would follow Grand Cutoff Bayou, the vulnerable spot at Widow Graham Bend that had concerned Ewens. At both places, the distance from the Mississippi across to the Red is about five miles. Using the cost figures for the Plaquemine Lock, which Thomas reckoned was built in "similar conditions," the upriver canal and lock would run more than six times the estimate for the dam, putting the total bill for the entire project at close to $3 million.[48]

◆ ◆ ◆

"PRACTICABLE, BUT NOT URGENT" is how the Mississippi River Commission concluded its November 21, 1913, report, settling for a time the question about damming Old River. In the three years since Thomas's 1910 fieldwork, the projected cost for the dam, canal and lock, and connecting levees had swollen to over $12 million, quite an eyebrow-raising figure at the time. The commission backed away from that potential boondoggle by rationalizing that the three rivers junction would gradually close by itself if the Corps ceased its dredging operation there. With absolute faith in Eads's Confinement Theory, the commission's report explained how, over time while Old River closes naturally, the Mississippi would "adjust its cross section to the volume it carries." For downriver interests, the commission saw a gradual closure as a safer option than a more abrupt severance by damming the outlet, which could raise the maximum flood height as much as four feet, more than enough to overtop the levee at New Orleans. Though the Old River dam was off the table, the commission recommended construction of the canal and lock at the Union Point site. If Old River is allowed to close itself, the commission agreed with the antiseparation proponents that the Plaquemine Lock was too far south to make commerce feasible between the Atchafalaya/Red/Ouachita/Tensas region and upper Mississippi ports.[49]

Since the Mississippi River Commission's formation in 1879, Old River had dominated its deliberations, public hearings, and fact-finding efforts. Always at the heart of the issue lay the seemingly irreconcilable navigational and flood control concerns of two opposing factions: the residents of the "interior basins" (Atchafalaya/Red/Ouachita/Tensas) and the people living along the Mississippi below Old River. The 1913 report placed a higher priority on the protection and safety of New Orleans and the other downriver communities, setting a precedent for future decisions about controlling the Mississippi River. By not ruling out the closure of Old River by natural processes, the commission held on to the option of flood relief and economic development for the interior basins. Indeed, the partnering-with-nature tone of the report beautifully justified maintaining the status quo, postponing indefinitely any action on the divorcement issue.[50]

For its part, the Mississippi kept spawning floods every two or three years. The back-to-back overflows of 1912 and 1913 caused nearly $80 million in damages throughout the Lower Mississippi Valley. After a two-year respite, the great river lifted again and swarmed south from February to April 1916. The gage at New Orleans registered the fifth highest crest on record. That same year, an important US Supreme Court decision swept away the legal objections to using federal funds to protect private property with levees. At the heart of the case was Eads's Confinement Theory.[51]

◆ ◆ ◆

IN HIS BOOK *RISING TIDE* (1997), John Barry exposes the flawed connection between the levees-only doctrine and Eads's idea about confining the river to make it deepen its bed. The distinction is important because the mistaken association of levees with river confinement serves as the foundation for the Supreme Court's ruling in the case known as *John F. Cubbins, Appt. v. Mississippi River Commission and the Yazoo-Mississippi Delta Levee Board*. Cubbins, a northwest Mississippi landowner, sued the Mississippi River Commission and the local levee board for building levees that elevated the height of the river, causing it to overflow and flood his land. He argued that, before the levees closed off natural outlets, the river spread out into adjacent basins and then drained harmlessly by way of various streams down to the Gulf. The Court stated that Congress created and funded the commission to "improve the river by building levees along the banks in order to confine the waters of the river within its natural banks, and, by increasing the volume of the water, to improve the navigable capacity of the river."[52]

Barry points out what should have been obvious. Levees are routinely placed well back from the river banks and cannot confine and concentrate the river's flow enough to cause scouring. When Eads talked of confining the river, he envisioned the use of jetties, which would extend out into the Mississippi and constrict the current. Humphreys and Abbott, who proposed the levees-only plan back in 1867, advocated levees, not as a means to deepen the river, but to wall off the cultivated and settled flood plain from the Mississippi during periodic flooding. In fact, Humphreys and Abbott disagreed strongly with Eads, maintaining that much of the Mississippi's bed is composed of hard clay and is resistant to scouring.[53]

As the Mississippi River Commission and the Corps of Engineers placed increasing emphasis on levees in the late 1800s, the flood protection rationale became blurred. By the time that the Supreme Court considered the constitutionality of levees, the justices associated the walls of earth crouching as far as a mile away from the active channel with improving the river's navigation, as opposed to protecting private property. However misguided it may have been, *Cubbins v. Mississippi River Commission* helped legitimize Congress's acceptance of a wider federal responsibility for controlling the Mississippi. After the 1916 flood, partnerships and compromises in Congress, bundling together the need for higher Mississippi River levees with projects in the West, succeeded in earmarking $45 million for the commission's use. Although the onset of World War I cut that funding short, the people of the Lower Mississippi Valley could now look beyond their own communities and states for congressional action on flood protection.[54]

During those war years, the Mississippi was uncharacteristically stable. Five years passed before the next big overflow in 1922. After that, four long years drifted by while the river gently rose and fell within its banks. In river towns like Greenville and Lake Providence,

New Roads and New Orleans, and the Atchafalaya Basin communities of Simmesport and Melville on down to Morgan City, folks slept easy on spring nights as the Mississippi kept a reassuring distance from its levees.[55]

At Old River, the restless forces set in motion by Shreve's cutoff and the removal of the Atchafalaya raft continued to seek a state of equilibrium. E. J. Thomas's work there in 1910 and 1911 showed that the Atchafalaya's channel was growing deeper and wider while the Mississippi's bed adjacent to Old River was rising.[56] Over the next twenty years, the occasional direction reversal of the water flowing in Old River would happen less frequently and people born at Red River Landing during 1882, the year Ewens measured the flood and the three rivers, would live to witness the end of that phenomenon.

While Old River was evolving, the Mississippi River Commission's strategy for the management of the Mississippi and tributaries rested comfortably and complacently behind the sinewy mounds of dirt standing guard along the river. After all, since the commission's inception, levees built to its specifications had never failed.[57] When the harbingering storms of late 1926 began to blow history's biggest flood into the Mississippi Valley, the twentieth-century mound builders' faith in their earthworks remained unshakable. Part of the chief of engineers' report to the secretary of war in 1926 might have been lifted directly out of the previous century's Humphreys and Abbott document: "Levees afford the only practicable means for flood control" as opposed to "reservoirs, cut-offs, and outlets."[58]

CHAPTER 4

The Other Side of the Flood

When trying to imagine what might happen if the Mississippi River jumps its channel and moves into the Atchafalaya Basin, the floods of 1927 offer the only comparable historical experience. The succession of storms and overflows during that killing spring swelled the Mississippi's volume to around 2.4 million cubic feet per second.[1] Roughly half of that deluge diverted from the main channel into the Atchafalaya. Should the channel jump occur, it will likely happen at the height of a similar flood. Although only a portion of the Mississippi diverted in 1927, the destructive force loosed on the Atchafalaya Basin provides a chilling look at a possible future scenario.[2]

The levees meant to contain that flood were a false hope. On the Lower Arkansas River, the grinding waters chewed through the earthen dikes lining its south bank on April 21, releasing part of the flood into the upper Bayou Boeuf drainage (see Figure 4.1). This natural flume through the floodplain carried the renegade water southward along the dry side of the Mississippi River's mainline levees from southeast Arkansas into northeast Louisiana. Days later, major levee crevasses on the west side of the Mississippi between Tallulah and Old River pulled the heart of the flood out of the main channel, where it joined the water moving south from the Arkansas River. Louisiana towns like Jonesville, twenty-three miles away from the Mississippi, were squarely in the path of the new river surging down the Tensas Valley. At the height of the disaster, the water in the streets of Jonesville ran fifteen feet deep.[3]

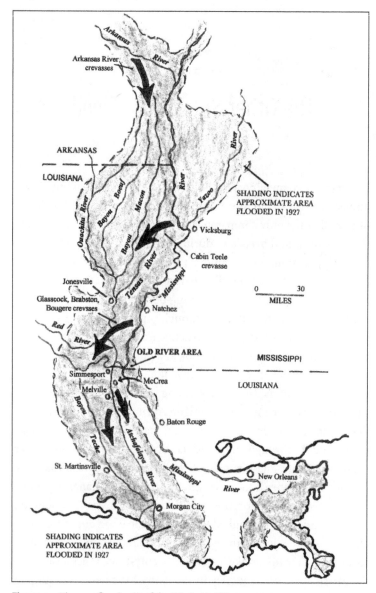

Figure 4.1 The 1927 flood west of the Mississippi River.

Forty miles south of Jonesville and fifteen miles to the west of Old River, the levees along Bayou des Glaises, at the head of the Atchafalaya Basin, extended across the route of the redirected flood. Designed to control a small, bottomland stream, these modest earthworks became a fragile line of defense against an advancing sea fed by the crevasses to the north and the weight of the Red River and its tributaries. Despite the sandbagging efforts of an army of laborers, the Bayou des Glaises levee between Hamburg and Simmesport melted away on May 12. Down below, only the towns and farms along the Atchafalaya River stood between the flood and the Gulf of Mexico.[4]

Partitioned from the inland flood by its levees, the Atchafalaya River grew black and menacing, carrying over six hundred thousand cubic feet per second from the Red River combined with the Mississippi tide rolling through Old River.[5] On May 17, the west levee at Melville broke, and upstream at McCrea the Atchafalaya's east levee crevassed (see Figure 4.1). With these levee failures, the separate floods attacking the Atchafalaya Basin joined forces in a southbound juggernaut covering twenty miles a day. Unstoppable now, the wall of water burned through the towns of Arnaudville, Breaux Bridge, St. Martinsville, New Iberia, Jeanerette, Franklin, and Morgan City. Photographs of Melville show water lapping at the eaves of houses. Farther south, the flood stood up to twelve feet deep over much of St. Martin, Lafourche, Terrebonne, St. Mary, and Assumption parishes. Some sixty thousand refugees huddled on rooftops, drifting in fishing boats, and gathered on Indian mounds looked out on the unimaginable. A vast new lake covering over five thousand square miles reflected the sun and stars back into space.[6]

Over on the Mississippi River, the Old River outlet and crevasses at places like Mound Landing, Cabin Teele, Glasscock, Brabston, and Bougere significantly lowered the flood's crest. Out of danger, New Orleans remained safe behind its levee. Ironically, it was dynamite and not the Mississippi River that caused the great crevasse thirteen miles below the Crescent City at the community of

Caernarvon. The blast came after two weeks of furtive deliberations that included New Orleans business leaders, the city's mayor, the governor of Louisiana, members of Congress, the chief of engineers, and the Mississippi River Commission. Despite the protests of some engineers who doubted the danger facing New Orleans, the destruction of the Caernarvon levee sacrificed thousands of homes in St. Bernard and Plaquemines parishes. Further downriver, the new Point a la Hatche flood-control outlet diverted part of the floodwater into Breton Sound.[7]

Until then, the levees-only strategy advocated by Humphreys and Abbott, the Civil War–era engineers, had dominated the Corps of Engineers' flood-control policy. The 1927 floods ushered in a new age as stunned US Army engineers watched the levees dissolve and pull the murderous floodwaters out of the Mississippi's channel. In the overflow spreading out across towns and farms, the Corps glimpsed the future of flood control. After 1927, the Mississippi River channel would have to include places where intentional crevasses mimic those that nature had rent in the mainline levees. These artificial crevasses, confidently christened "floodways" and "spillways," could be triggered at strategic moments during future floods to diffuse the Mississippi's force and dramatically reduce the threat to New Orleans.[8]

◆ ◆ ◆

THE 1928 FLOOD CONTROL ACT embraced a floodways plan based upon designs hurriedly prepared by the Corps of Engineers and the Mississippi River Commission within months after the floodwaters receded. Major General Edgar Jadwin, the Corps' chief of engineers, presented his version of the new floodways strategy to President Calvin Coolidge and Congress in December 1927.[9] A high priority for the sixty-two-year-old general, a decorated veteran of the Spanish American War and World War I, was an artificial outlet near New Orleans. After considering several possible locations,

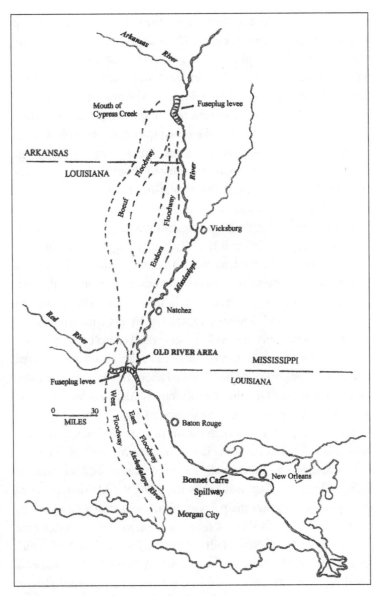

Figure 4.2 Lower Mississippi Valley floodway plans: 1928–1938.

Jadwin selected a spot upstream from the Crescent City at a sharp turn in the river known as Bonnet Carré Bend (see Figure 4.2). Here the Mississippi's current pounded against the east bank at a vulnerable stretch along a six-mile-wide neck of bottomland separating the river from Lake Pontchartrain. Since the mid-nineteenth century, Bonnet Carré Bend had seen several spectacular crevasses that pushed floodwaters into the lake across the intervening plantations and bayous.[10] In the steamboat era, the church steeple at the village of Bonnet Carré (square bonnet), situated about four miles upriver from Bonnet Carré Bend at the present-day town of Edgard, was a landmark for river travelers.[11]

Humphreys and Abbott were against creating an outlet at Bonnet Carré Bend, fearing that the Mississippi might turn all or most of its flow into Lake Pontchartrain. The two engineers also argued that the accumulation of the river's sediment brought in through the outlet would impair navigation in the shallow lake. In Jadwin's plan, Bonnet Carré Spillway, completed in 1931, incorporated a structure seven thousand feet long comprised of 350 bays, each containing twenty creosote timbers called needles raised and lowered individually by gantry cranes. Located about thirty-two river miles upriver from New Orleans, the spillway could, it was hoped, divert just enough water to protect the Crescent City.[12]

Much like New Orleans, the town of Cairo, Illinois, some 960 river miles to the north at the junction of the Mississippi and Ohio Rivers, occupies a commercially lucrative location and is extremely vulnerable to flooding. Just below the town, the Mississippi's constricted channel slows the passage of floodwater welling up at the confluence of the two rivers. Jadwin's solution to Cairo's dilemma was the Birds Point-New Madrid Floodway in southeast Missouri. In contrast to the gated spillway down at Bonnet Carré, Jadwin's floodway opposite Cairo would be activated by a special type of levee designed to fail when the river reaches a predetermined stage. In the parlance of the Corps of Engineers, a levee built to break under certain conditions is called a fuseplug. As its name implies, fuseplug levees will be dynamited if they resist the floodwaters they are engineered to admit. Of course, fuseplug levees are much cheaper to construct and maintain than gated control structures. On the negative side, once the fuseplug levee collapses, the water coming into the floodway won't stop until the flood subsides. Not surprisingly, the people living in the proposed floodway were appalled to learn that federal officials wanted to sacrifice their homes and 130,000 acres of farmland to keep Cairo safe and dry.[13]

Between Jadwin's two relatively small artificial outlets at Cairo and New Orleans, the chief of engineers' plan commandeered two vast natural drainage systems he considered essential to his floodways strategy. Together, the Boeuf and Atchafalaya floodways would capture half of the next superflood.

The head of the Boeuf Floodway was the mouth of Cypress Creek on the Mississippi's west bank a few miles below the mouth of the Arkansas River (see Figure 4.2). Before the levee was built to close it, a natural outlet for the Mississippi at this location pulled overflow water from the main channel into the Boeuf and Tensas drainage basins. Farmers down below Cypress Creek suffered increasing losses during the big floods of 1912, 1913, and 1916. In the flood of 1916, around 336,000 cubic feet of Mississippi River water per second came through the Cypress Creek outlet, wreaking havoc all the way down into the Atchafalaya Basin. Landowners in the flooded areas appealed to the Mississippi River Commission. Levee districts in southeast Arkansas and northeast Louisiana received permission to extend the mainline levee across the Cypress Creek gap in 1921. Predictably, denying the Mississippi its Cypress Creek outlet meant higher flows and damages downstream during subsequent floods.[14]

Under Jadwin's congressionally approved Boeuf Floodway, the Cypress Creek outlet would be resurrected. The chief of engineers' plan was audacious, appropriating over two thousand square miles of prime delta farmland, an area roughly equal to the state of Delaware. Angling away from the Mississippi River in a southwesterly direction, the floodway would follow Bayou Boeuf down through the Tensas Basin. While the rest of the mainline levees along the Mississippi would be enlarged and strengthened, the opening to the Boeuf Floodway, some thirty miles wide, would retain its pre-1927 levee to function as a fuseplug.[15]

Like the people in the Birds Point-New Madrid Floodway zone, the folks in the Boeuf/Tensas drainage were being told to sacrifice their homes and property for their Mississippi neighbors on the other side of the river, whose farms and towns went under during the great April 1927 crevasse at Mound Landing, near Greenville. If built, Jadwin's Boeuf Floodway would protect these people and their property during the next big flood. Although Jadwin considered placing a floodway in Mississippi, he ruled out the option because the line of hills forming the eastern side of the Yazoo Basin would simply channel the floodwater right back into the Mississippi River.[16]

Jadwin's plan also called for fuseplug levees at the head of the other principal "safety valve" (the chief of engineers' term) above New Orleans, the Atchafalaya Floodway. Here, two fuseplug levees totaling about seventy miles in length defined the upper end of East and West floodways, occupying the land on either side of the Atchafalaya River (see Figure 4.2). The Atchafalaya River's repaired levee system would separate the stream from the two floodways, and the entire floodway system would be enclosed by outside guide levees. The Atchafalaya River levees end about sixty-five miles below the Old River junction. There the river and the two floodways would combine to become a single broad floodway leading down to Atchafalaya Bay and the Gulf of Mexico.[17]

The Atchafalaya Floodway played a critical, double roll in Jadwin's flood-control scheme. In addition to carrying the floodwaters from the Red River system and the portion of the Mississippi coming through the Old River outlet, the floodway would also serve as an extension of the Boeuf Floodway, receiving the floodwater coming from the Cypress Creek fuseplug levee. By redirecting this much water out of the Mississippi River, the combined floodways would bring the 1927 flood back to the Atchafalaya Basin, which Jadwin dismissed in his report as "largely swamp land."[18]

Major General Jadwin died in 1931, before any of the floodway construction projects he initiated became operational. Although not all of his floodways were built, some stand today much as he designed them, complete with their fuseplug levees. Two of the floodways designed under Jadwin's supervision, Bonnet Carré and Birds Point-New Madrid, have been used as recently as 2016 and 2011 respectively, but operation of the Corps of Engineers' entire Mississippi River floodway system awaits an event hypothesized in Jadwin's plan, a doomsday augury known as the "project design flood."[19]

After the largest Mississippi River flood on record, few doubted that floods of equal size or larger would rise again. In order to design an adequate plan to cope with another big flood, the engineers needed a hypothetical event with specific, analytical dimensions. To give the leviathan form and substance, the US Weather Bureau (forerunner of the National Weather Service) estimated the "maximum possible" Mississippi Valley flood based upon available data on storm patterns. At the same time, the Mississippi River Commission used measurements from the floods of 1913 and 1927 to come up with its own statistical prediction of the "maximum probable" flood. Jadwin's team combined these two projections and the chief of engineers' 1927 plan debuted the concept of the project design flood, or simply the "project flood," with a volume of three million cubic feet per second, projected to be about 20–25 percent larger than the Mississippi at the height of the 1927 floods. This imagined superflood remains the yardstick for Lower Mississippi Valley flood control.[20]

◆ ◆ ◆

WORD OF THE BOEUF FLOODWAY PLAN spread through the farming communities of southeast Arkansas and northeast Louisiana like an epidemic. Less than a week after Jadwin delivered his flood-control plan to Congress, members of levee boards and concerned landowners in the proposed floodway zone assembled in Monroe, Louisiana, to organize in opposition. Speaking with a unified voice, the group swiftly adopted a resolution denouncing the floodway. Though not as well organized as the opponents of the Boeuf Floodway, landowners in the Atchafalaya Basin raised the same objections. To these taxpaying American citizens, the federal government's intention to seize control of their land was unacceptable, as was Jadwin's plan to activate the floodways with fuseplug levees, which could not be regulated once the levees crevassed. In a separate critique of the chief of engineers' flood-control strategy, Louisiana's Board of State Engineers also objected strongly to the use of fuseplugs.[21]

During the months following the passage of the 1928 Flood Control Act, the residents of the proposed floodways experienced confusion, fear, and resentment as stakeholders tried to sort out the new law's implications. Cobbled together in haste, the bill encouraged brawling over bureaucratic turf by authorizing a flood-control project "in accordance with" Jadwin's plan while also directing the chief of engineers to serve on a board to consider alternative flood-control measures submitted by the Mississippi River Commission. Jadwin was hardly impartial on this issue. In his own plan, he recommended abolishing the Mississippi River Commission's shared civilian/military governance, giving the chief of engineers sole authority to "plan and direct the work on the Mississippi River."[22] For the conservative Coolidge administration, the $296 million estimated cost for Jadwin's plan was far more palatable than the Mississippi River Commission plan's estimated $775 million price tag. A significant part of the commission plan's higher cost reflected a sharp difference of opinion on the critical issue of federal payment to landowners for the use of their property, referred to as "flowage rights."[23]

Jadwin was adamant that federal dollars must not be awarded to landowners in payment for flowage rights in the Boeuf and Atchafalaya floodways. Under his plan, the states of Arkansas and

Louisiana, with their local levee boards, would have to bear flowage costs and furnish rights-of-way for all flood-control works. In addition, states would be required to pay 20 percent of the construction cost for the two floodways' guide levees. As for the towns within the floodways—Morgan City, Melville, and Simmesport—Jadwin prescribed encircling ring levees; however, these rural communities would be expected to cover 50 percent of the cost of this protection. The chief of engineers justified his draconian measures by arguing that these two floodways were already part of what he called the "natural flood bed of the river." In Jadwin's view, the federal government had no business paying landowners whose property would likely be flooded intermittently anyway. In response to suggestions that the Corps of Engineers should purchase the Boeuf and Atchafalaya lands outright, Jadwin countered that the floodways will be used so infrequently that federal ownership would be impractical.[24]

♦ ♦ ♦

WHILE FLOODWAYS AND HIGHER LEVEES were going to be essential to the future of flood control, these were passive works that might stand unused for years. Brigadier General Harley B. Ferguson wanted to wade in and grab hold of the Mississippi River to tame some of the wildness out of it. After all, hadn't American engineers chiseled through a mountain with dynamite, dinosaur-sized steam shovels, and dredges to make the Panama Canal? With such earth-shaping muscle at his disposal, Ferguson reasoned, even a force as powerful as the Mississippi can be brought under control.[25]

Aside from Ulysses S. Grant's failed attempt during the Civil War, no one had seriously proposed cutting off the Mississippi's meandering loops since the Old River cutoffs at Turnbull and Raccourci bends in the early nineteenth century. Before Henry Shreve's precipitous attack on Turnbull Bend, the general consensus held that shortening the Mississippi River was a harmless improvement on nature.[26] That attitude changed when Shreve's cutoff and the follow-up cutoff at Raccourci Bend revealed the capricious and hazardous consequences of trifling with a river the size of the Mississippi. In their influential Mississippi River study of 1861, Humphreys and Abbott argued against making cutoffs, warning of two dangers: flooding below the cut and, because the artificial channel would not be as deep as the natural one, perilously high water at the site of the cutoff. Even in the rush to employ a battery of flood-control measures after 1927, most of the authorities in the Corps of Engineers and Mississippi River Commission still considered artificial cutoffs to be too unpredictable.[27]

After his appointment as president of the Mississippi River Commission in 1932, General Ferguson quickly effected a radical shift in flood-control strategy. By that time, the lanky, pipe-smoking engineer from North Carolina had established an enviable career record, including famously leading the team that raised the battleship USS Maine in Havana Harbor. Although Ferguson was new to the Mississippi Valley, the 1927 floods had captured his interest, and the maps and reports he studied convinced him that the big river needed straightening. To skeptics of cutoffs, Ferguson characterized the Mississippi River as a stream confined by natural features like bluffs and incorrectly sited human-made revetments. Unable to adjust itself in a restricted floodplain, several big meander bends had become, in Ferguson's words, "obstacles to an orderly regimen of the river." They slowed down the Mississippi's floods, causing them to pool and spread out. Ferguson reasoned that this dangerous high water should be hurried along to the Gulf as quickly as possible. His solution would be a carefully sculpted channel with certain meander bends cut away and precise dredging and revetment placement in the reaches between the cutoffs.[28]

Reminiscent of James B. Eads's nineteenth-century Confinement Theory, Ferguson believed that the Corps could "direct the river's energy in the improvement of the channel." By accelerating the current through a series of cutoffs, the river's natural power would dredge its own bed and deepen its channel. To quote the new Mississippi River Commission president, "If you prevent [the Mississippi] from working the bends then it has to work on the bottom."[29] As it had in Eads's time, the idea of harnessing the Mississippi to solve the flooding problem sounded logical, but would it work? By the early 1930s, testing conducted at the Corps' new Waterways Experiment Station at Vicksburg and an independent study by the Board of Rivers and Harbors (of which Ferguson was a member before moving to the Mississippi River Commission) indicated that the cutoff approach had merit. Despite the misgivings of some of his colleagues, Ferguson's extreme confidence in this strategy helped assuage concerns that the Mississippi River's behavior just might run counter to the laboratory models. There was only one way to find out.[30]

In late 1932, draglines and dredges started digging across the necks of meander bends between Old River and the mouth of the Arkansas. Henry Shreve would have been impressed by the machines the Corps marshalled for the cutoff work. Looking like floating construction sites, the steam-powered dredges carried giant cranes that lowered an arm encasing a suction tube to depths up to ninety feet. The biggest dredges wielded suction arms equipped with enormous cutterheads resembling something out of Jules Verne's imagination: a whirling, steam-driven battering ram wrapped in spirals of sediment-devouring teeth. Barges decked with vacuum pumps clustered around the dredges to inhale the dislodged material and wash it down floating pipelines to dumping sites on shore or to places where the sand and clay could form submerged dikes to help direct the river current.[31]

The distances across the necks of the meander bends varied from just over a mile to almost five miles. At each one, timber was cleared away, and a squadron of dragline cranes with scraper buckets set to work removing the overburden along the line of the cut. For the channel excavation, Ferguson deployed dredges at the upper and lower ends of the cutoff to progress toward the middle of the neck, bringing the river with them. When the channel was ready, the crew dynamited the narrow partition of earth left standing between the two halves of the trench. Most of the cutoffs required several years of dredging to fully develop their channels. In 1940, only two of the thirteen cutoffs Ferguson ordered between 1932 and 1939 were deep enough to carry the Mississippi's full discharge in high water.[32]

During Ferguson's tenure, his engineers dug twelve cutoffs below the mouth of the Arkansas River. The Corps also completed a natural cutoff that began forming at Yucatan Bend below Vicksburg in 1929. After Ferguson's retirement in 1939, the Corps finished three more cutoffs between the mouth of the Arkansas River and Memphis.[33] The bold engineering initiative shortened the Mississippi River by about 170 miles, and the big 1937 flood seemed to validate Ferguson's prediction; flood heights gaged 2.5–3.5 feet lower than expected through the Memphis to Old River reach. Along with the lowered flood height, Ferguson estimated that the cutoffs and channel sculpting increased the river's carrying capacity by half a million cubic feet per second.[34] But what about the long-term effects of abruptly severing fifteen of the Mississippi's meander bends (sixteen counting the natural cutoff at Yucatan)? Such a quick victory over a river that counts time in tens of centuries turned out to be an illusion.

Even before all of the cutoffs were finished, Ferguson was aware of one negative repercussion occurring at the latitude of Old River and below. As predicted, the accelerated current through the new cutoffs scoured the riverbed, sending tons of disturbed sediment downstream. At the lower end of the Mississippi, the slope to the Gulf flattens out, slowing the current enough to make the sand and silt settle and aggrade or raise the bed. With the riverbed accumulating sediment, the floodwaters from this point down to New Orleans rose higher.[35] Later studies confirmed this trend and pointed to a larger concern. In 2000, river engineers David S. Biedenharn, Colin R. Thorne, and Chester C. Watson reported that the Mississippi had been a "broadly stable system" before the 1930s cutoffs, and that sixty years later the river had not recovered from Ferguson's radical surgery. As the river readjusts, it is gaining back some of its lost length. Historian John Barry suggests that as much as one-third of the amputated mileage has returned. Examination of maps from the 1960s to the 1990s shows that, where the Mississippi still has some wiggle room, its channel is in continuous flux. Likewise, the initial lowering of the river's level between the Arkansas River and Old River was only temporary; the dangerously high flows through this reach were back by the early 1960s.[36]

Ferguson's battle with the cutoffs was only part of the war. In 1932, a flotilla of the Corps' dredges left the Mississippi through the Old River junction and steamed down the Atchafalaya. So began the mission to transform over twelve hundred square miles of sluggish bayous and delta farmland into the world's largest floodway, the great trough that would carry half of the fabled project flood to the Gulf. The floodway initiative was a complete reversal of the Corps' late-nineteenth-century strategy to diminish the Atchafalaya Basin's capacity to capture the Red River and separate it from the Mississippi. To play its part in the floodway design, the Atchafalaya River would need to triple its maximum volume from two hundred thousand to around six hundred thousand cubic feet per second.[37]

Like the cutoffs altering the Mississippi's main channel, the dredging in the Atchafalaya Basin was unprecedented. At least the Corps' laboratory models had supported the work on the main channel. Similar testing for dredging in bayou country was just beginning, so Ferguson and his engineers referred to the Atchafalaya channel-digging as "experimental." The Lower Atchafalaya River proved to be especially difficult. About fifty miles below Old River the Atchafalaya's slope flattens, and sediment traveling in the current begins to settle out. Over the remaining seventy miles down to the Gulf, the water filters slowly through a maze of bayous and lakes shaped and reshaped by the silt. At about the latitude of Morgan City, some fifteen miles above Atchafalaya Bay, the Corps' dredges had to cut through a dense clay ridge, a remnant from an ancient Red River channel. The clay impeded both the enlargement of the existing Atchafalaya River channel and excavation of the floodway's second Gulf outlet channel through Wax Lake, about ten miles west of Morgan City.[38]

Once started, the dredging in the Atchafalaya Basin became a never-ending chore. Dredges had to repeatedly go back and rework channels clogged by shifting sediment. Corps of Engineers historian Martin Reuss likened the Lower Atchafalaya to "an unruly child [refusing] to behave as the adults wished."[39] Throughout the 1930s, a growing fleet of dredges labored in the Atchafalaya Basin, desperately trying to fashion a channel capable of carrying the floodway's share of the project flood.[40] If the swamps stubbornly resisted Ferguson's channeling, at least the Corps of Engineers enjoyed the approval of most of the people in whose midst they were working. Transcripts of congressional hearings held during the Depression years show plenty of local support for the dredging.[41]

Most of the people in the Atchafalaya Basin also looked favorably upon the guide levees the Corps constructed in tandem with the dredging. These levees forming the boundaries of the floodway would keep the high water away from many areas that were inundated in 1927 and protect thousands of acres of south Louisiana sugarcane and rice fields. For those who owned good farmland inside the floodway, however, the hard times quickly became much worse. Land values plummeted as much as 80 percent. Banks began denying loans to people in the floodway. Those with existing loans faced collection notices from nervous creditors. In early 1934, H. P. Mounger, a prominent farmer at the northern end of the East Floodway, told a congressional committee that the land he had farmed most of his life had become a "liability instead of an asset" because of the floodway plan. Mounger represented a group of farmers and landowners who wanted the north end of Pointe Coupée Parish left out of the floodway. Besides stressing the value of the farmland, he reminded the members of congress that General (retired) John A. Lejeune, famed World War I marine commander, was born and raised in the upper East Floodway, further noting (or

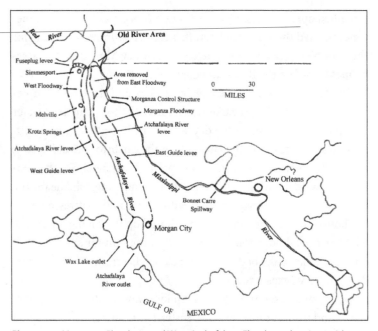

Figure 4.3 Morganza Floodway and West Atchafalaya Floodway showing guide levees. Adapted from Chatry, Trans. ASCE, 1961, 108 in Hebert, "The Flood Control Capabilities of the Atchafalaya Basin Floodway," Figures 1 and 2.

perhaps warning) that the graves of the general's parents were there. Testimony from Mounger and other residents of the area helped convince the Corps of Engineers to begin drafting plans to move the opening of the East Atchafalaya Floodway in order to exclude over six thousand acres in the north end (see Figure 4.3). Like Bonnet Carré Spillway, the new design would take advantage of one of the Mississippi's notorious crevasse sites, a bend in the river about twenty miles south of Old River at the town of Morganza.[42]

During the nineteenth century, the Morganza levee crevassed at least eight times. Here, the Mississippi wheels around a sharp western turn called Morgan's Bend and leans hard against the west bank. The worst break occurred in the 1867 flood, when the river tore a prodigious mile-wide gap in the mainline levee. A crevasse in 1874 remained open for nine years. Although Morganza residents some-times blamed the Raccourci cutoff, fifteen river miles upstream, for their problems, two of the nineteenth-century Morganza crevasses happened before the Raccourci channel cut.[43]

The Morganza levee failed again in 1890, and the land for miles above and below the crevasse was inundated. It was the backwater lapping for hours against the dry side of the levee that caused a fif-ty-foot-long section on the land side of the earthwork to collapse at the town of New Texas, three miles upriver. Fortunately for the peo-ple of New Texas, Mrs. E. F. Oubre took action when she heard the 1:00 am cry of "crevasse." Armed with sheets and mattresses from her house, Mrs. Oubre and her son shored up the caving levee and, with their neighbors, fought the flood in the rain for some sixteen hours. Thanks to the bedding and sacks of dirt from the Oubre gar-den, the New Texas levee held.[44]

Once completed, the Morganza Floodway would protect Mor-gan's Bend from future crevasses. Instead of heading the redesigned floodway with a fuseplug levee, the Corps wisely opted for a gated control structure just north of the town of Morganza. While the Morganza Floodway plan rescued H. P. Mounger and rest of the landowners at the upper end of the East Floodway (see Figure 4.3), diminished property values still plagued their neighbors over in the West Floodway. Residents there were also justifiably concerned about the fuseplug levee stretching from Hamburg to Simmesport (see Figure 4.3). Some comfort may have come from Corps reports to Congress that the redesigned East Floodway incorporating the Morganza control structure would reduce (but not eliminate) the need for the West Atchafalaya Floodway.[45] On paper, the Morganza Floodway and the Atchafalaya River could potentially carry 1.2 million cubic feet per second, 80 percent of the Atchafalaya Basin's share of the project flood.[46]

The work in the Morganza and West floodways along with the improvements to the carrying capacity of the Atchafalaya River became critically important after the people of northeast Louisiana and southeast Arkansas successfully thwarted the Corps' plans for the Boeuf Floodway and its downsized alternative, the Eudora Floodway (see Figure 4.2). The Boeuf/Eudora outlet lost political support after Ferguson's cutoffs showed promise in lowering floods, and the Mississippi River Commission approved placing reservoirs on the Red, Arkansas, White, and Yazoo drainages. With the Boeuf/Eudora outlet scrubbed, the Atchafalaya Basin assumed the major role in Mississippi River flood control.[47] The Corps and the Mississippi River Commission called their post-1927 multifaceted approach to flood control "The Mississippi River and Tributaries System," a so-far-successful paradigm that remains viable in the twenty-first century.[48]

◆ ◆ ◆

OLD RIVER WAS A DARK HORSE during the 1930s and early 1940s. While the great debate that Martin Reuss called the "Battle Over Floodways" roiled,[49] a lonely dredge plied Old River's reach between the Mississippi and the Red, doggedly keeping the passageway navigable between the rivers. Although the Old River connection seemed stable, with its occasional reversal in the direction of flow, ominous changes were creeping into the hydrology of the old confluence. By 1938, the Corps' dredging in the Atchafalaya River had ramped up the distributary's carrying capacity to an estimated five hundred thousand cubic feet per second. Over on the Mississippi side of Old River, Ferguson's upstream cutoffs altered the river's dynamics considerably, pulling the main current away from the Angola bar on the east bank and throwing it against the Old River side. As a result, increasing amounts of water left the main channel during high flows and poured down the Atchafalaya.[50]

Keenly focused on preparing the Lower Mississippi Valley for the project flood, the Corps' ambitious campaign of dredging and cutoffs overlooked the subtle transformation occurring at Old River. In 1944, the Corps' dredges severed Carr's Point, the lengthening

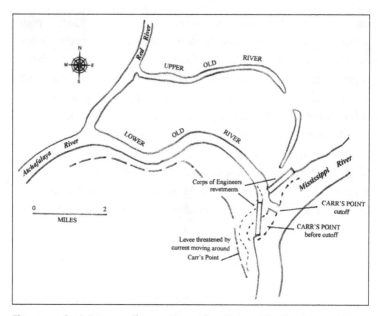

Figure 4.4 Carr's Point cutoff, 1944. Adapted from Hebert, "The Flood Control Capabilities of the Atchafalaya Basin Floodway," Figure 8.

Mississippi River sandbar described in Chapter 3 that threatened to close Old River. The movement of water around the south end of Carr's Point bar was encroaching on the mainline levee just south of Old River. The cutoff operation radically altered the angle of the Mississippi's connection to Old River's lower arm, making a more direct channel for the Mississippi's contribution to the Atchafalaya via lower Old River (see Figure 4.4). As this westward flow became more intense, scouring in the upper Atchafalaya River slowly increased the distributary's carrying capacity.[51]

In March 1945, the Red River raised one of its largest floods on record. It crested higher than the Mississippi, so that gravity pulled the Red's floodwater eastward through Old River, causing the highest stages since 1927 on the Mississippi River gages south of Red River Landing. Desperate residents of the Red River Backwater area called on the Corps to open the Morganza Floodway in hopes that the new outlet might relieve their flooding, but the Morganza control structure was far from operational due to World War II construction delays. Although the Corps did consider dynamiting the mainline levee at Morganza, the army engineers decided that a man-made crevasse there would have little or no effect on the water level up on the Red River. To save New Orleans, the Corps opened the Bonnet Carré Spillway for the first time since the 1937 flood.[52]

When the 1945 flood finally began to recede in May, the direction of the current reversed in Old River, once again carrying Mississippi water toward the Atchafalaya. Corps of Engineers personnel took note, but no one at the time realized that an event of historic significance had occurred. Old River's century of indecision ended quietly while the United States celebrated the victories of its armed forces in Europe and the Pacific. After that spring, the current would only flow in one direction through the seven-mile, dogleg reach. The push and pull between the Mississippi, Red, and Atchafalaya at Old River had ceased, ending the precarious balance achieved in the wake of Shreve's cutoff. The Mississippi River was moving into the Atchafalaya Basin.[53]

CHAPTER 5

Yet Trouble Came

Unlike his famous father, forty-seven-year-old Hans Albert Einstein would not have attracted attention on Vicksburg sidewalks in December 1951. In his plain, dark suit and barbershop haircut, he blended easily with the white-collar professionals of his day. One of the country's foremost authorities on river-borne sediment transport, Einstein was in the historic river town at the request of the Corps of Engineers and the Mississippi River Commission to help devise a solution to the emergency situation arising at Old River.

Given the critical nature of the problem, Einstein's presence in the Lower Mississippi Valley was inevitable. In the vicinity of Old River, the Mississippi was carrying anywhere from 218,000 to 1.5 million tons of sand, silt, and clay per day.[1] Part of this sediment load left the main channel and passed through Old River along with the Atchafalaya River's increasing share of the Mississippi. In planning for a control structure at Old River to halt the Mississippi's diversion, the Corps of Engineers needed to know how best to apportion the mud as well as the water.

The younger Einstein studied hydraulic engineering in Switzerland and immigrated with his family to the United States in 1938 as the likelihood of war in Europe increased. In 1947, he joined the faculty at the University of California at Berkeley, becoming a distinguished professor in the College of Engineering's highly regarded hydraulics program.[2] Although Einstein's consulting trip to Vicksburg in 1951 was his first experience with the Lower Mississippi River, it was not the engineer's first visit to the region; in 1941 his research for the US Soil Conservation Service brought him to Lafayette County, Mississippi, to study sediment transport in Goose Creek, a small stream just west of the university town of Oxford.[3]

The need for expert consultants of Einstein's stature stemmed from the Corps' own troubling conclusions about the changing dynamics at the three rivers confluence. In 1945, after publishing Harold N. Fisk's landmark geological study of the Lower Mississippi River, Mississippi River Commission president Brigadier General M. C. Tyler asked Fisk for his opinion on the likelihood of the Atchafalaya capturing all or most of the Mississippi's flow.

Fisk replied that such a scenario was a "definite possibility."[4] If Fisk was right, the Corps faced a complex dilemma. With the Morganza and West Atchafalaya floodways nearing completion, the Atchafalaya Basin was approaching readiness to meet its design purpose: to divert half of the Mississippi's biggest floods safely away from New Orleans and down the floodways to the Gulf. The ability to accomplish this goal hinged on improving the carrying capacity of the Atchafalaya River. But as the Corps' dredging and natural scouring increased the Atchafalaya's discharge capacity, the stream boosted its potential to swallow the whole Mississippi River.

Along with the bad news about the Mississippi and Atchafalaya Rivers, the Corps' Old River deliberations had to consider the plight of the farmers in the Red River Backwater. Since the nineteenth century, these landowners had called for the separation of the Red and Mississippi Rivers to reduce the flood threat to their homes and fields (see Chapter 3), caused when the Mississippi's high water spreads through Old River and backs up the Red and its tributaries. Their interests, of course, were always outweighed by Old River's usefulness as a navigation conduit and floodwater outlet. Now, the growing threat of the Mississippi's capture by the Atchafalaya cast the separation issue in a new light.

If a clean divorcement of the Red and Mississippi was unlikely, perhaps a control structure in Old River could reduce the flooding on the Lower Red River. In a 1945 Corps of Engineers report on the subject, Colonel Clement P. Lindner agreed with Fisk's opinion on the possibility of a catastrophic channel jump. Lindner also noted that, while protecting the farmland in the Red River Backwater wouldn't justify the expense to build and operate such a structure, prevention of the Mississippi's diversion into the Atchafalaya Basin would certainly be worth the cost. The beleaguered landowners in the Red River Backwater region could take heart that their protection would "become a by-product" (in Lindner's words) of the greater mission to keep the Mississippi flowing to New Orleans.[5]

◆ ◆ ◆

THE CRESCENT CITY'S WELL-BEING also depended upon the completion of the war-delayed Morganza Floodway. Lindner devoted a portion of his Red River Backwater report of 1945 to considering options for Morganza's all-important intake structure. Unlike the structure being discussed for Old River, which would be in continuous operation, the Morganza Floodway would remain closed most of the time. For comparative purposes, Lindner included a fuseplug levee as a low-budget option and considered a needle-type weir similar to that at Bonnet Carré. The most practical design, however, was a steel-gated structure built to carry the Texas & Pacific Railroad and Louisiana State Highway 1. Since both of these transportation routes had to cross the Morganza Floodway, economic and hydraulic considerations pointed to incorporating them into the structure's design instead of building separate trestles and bridges.[6]

During the planning for the Morganza intake structure, the Corps had an environmental issue to consider, something the engineers would have to face with increasing frequency in the latter part of the twentieth century. The 1945 flood forced the operation of the Bonnet Carré Spillway for only the second time since its construction (the first was the 1937 flood). After the flood, fishermen complained that the Mississippi water and sediment pouring into Lake

Pontchartrain through the spillway had damaged the oyster beds in Lake Borgne and the Mississippi Sound. Although the charge was of tenuous validity, the Corps felt the need to design the Morganza Floodway for frequent use in order to minimize the need to open Bonnet Carré.[7]

In his comparison of structure designs, Lindner pointed out that a fuseplug levee would have to be rebuilt every time the floodway is activated, which could occur as often as every three years. As for the needle-type structure, the consensus was that this kind of spillway would not pass enough water to handle larger floods, much less the project design flood. In order to play its role in managing the project flood, the Morganza structure should be able to draw as much as 650,000 cubic feet per second from the Mississippi River. To accomplish this, the gated structure being considered was around 4,000 feet in length with a weir crest or threshold elevation of 44 feet above mean sea level, the height at which the Mississippi would overtop its west bank at the fore bay or mouth of the spillway.[8]

Settling on the gated intake that incorporated the railroad and highway crossing, the Corps began construction in 1950 and completed the Morganza control structure in February 1953. The steel-reinforced concrete structure is 3,960 feet long, with 125 vertical lift gates operated by two twenty-five-ton gantry cranes.[9] Behind the floodgates, the Morganza Floodway extends southward approximately 20 miles and encompasses about 105 square miles. At its lower end, the floodway fuses with the Atchafalaya River and the West Atchafalaya Floodway (see Figure 4.3). Down at the Gulf of Mexico, the Corps equipped Morgan City with a 7-foot-high concrete floodwall in 1948 (raised an additional 10 feet in height in 1987). Once the Morganza Floodway was operational, some were concerned that repeated use of the outlet would introduce enough sediment to hamper its carrying capacity. Ironically, over on the Atchafalaya River the opposite process—rapidly increasing capacity—was forcing the Corps to move swiftly and catch the Mississippi River before it veered off course.[10]

◆ ◆ ◆

IN JUST EIGHT MONTHS, from October 1950 to May 1951, the Corps of Engineers produced one of its most important documents, a multidisciplinary report bearing a seventy-word title known simply by its main title, *The Atchafalaya River Study* (1951). The study confirmed with hard data the warnings voiced by Lindner and Fisk: capture of the Mississippi by the Atchafalaya was imminent. Based upon the Atchafalaya River's phenomenal growth between 1892 and 1950, the former distributary could become the Mississippi's main channel by 1968, carrying over one million cubic feet per second. Even more chilling was the news that the tipping point might be reached in 1960, after which the regime change would be rapid and irreversible.[11]

The group that generated this sobering information formed in the fall of 1950 at the behest of the chief of engineers, Lieutenant General Lewis A. Pick. Each team member was responsible for a different aspect of the problem. Fisk, who had recently left his post at Louisiana State University for a position with Humble Oil, headed the geological investigation. W. J. Turnbull at the Corps' Waterways Experiment Station in Vicksburg supervised the borings and their analysis. E. J. Williams, with the Mississippi River Commission, looked at how past engineering work had affected the Atchafalaya River. And Corps engineers R. A. Latimer and C. W. Schweizer coordinated the engineering study and prepared the final draft report.[12]

Comprehensive in scope, the report summarized the changing hydraulics at Old River, from Shreve and the removal of the Atchafalaya Raft through the perennial navigation problems in the passageway's upper and lower arms. According to Latimer and Schweizer, the first official notice about the Atchafalaya's possible capture of the Mississippi came in an 1852 report by Charles Ellet, the engineer whose mid-nineteenth-century flood-control plan was ignored in favor of the disastrous levees-only strategy (see Chapter 3). For his part, Ellet had correctly noted that his observation wasn't the first warning voiced about the Mississippi's move toward the Atchafalaya Basin. As mentioned in Chapter 2, visitors to the Old River area recognized the threat decades earlier.[13]

Good documentation dating from the 1880s tracks the persistent westward migration of the Mississippi's channel at Old River, a trend that seemed to be replicating the process that led to the formation of Turnbull Bend in the centuries before the coming of Europeans. The Corps' Carr's Point cutoff in 1944 hastened this channel shift to the west, shortening Old River's lower arm by about one mile and shifting the confluence of Old River and the Mississippi to the north by about the same distance. The Carr's Point cutoff probably also contributed to the sea change at Old River after the 1945 flood. Before 1944, according to Latimer and Schweizer's compilation of records, most of the flow through Old River ran from west to east, toward the Mississippi. As noted previously, this current flowed consistently westward after 1945, with the amount of water leaving the Mississippi steadily increasing from 17 percent to 23 percent in 1950.[14]

In addition to the changing dynamics at Old River, the team was concerned about the alarming trend in the twenty-mile reach of the Mississippi between Old River and Morganza (see Figure 5.1). In sharp contrast to the Atchafalaya River's expanding capacity, this short stretch of the Mississippi was decreasing its capacity due to sediment accumulation. As more and more of the Mississippi's flow diverted through Old River, the diminishing flow in the main channel was losing its ability to transport its sediment load, allowing the heavier silt and sand particles to settle on the bottom. The Mississippi's rising bed here posed a major threat to the Corps' flood-control scheme. In big floods, the Old River to Morganza reach needed to carry up to 2.1 million cubic feet per second. If that much water could not get through the reach, then more floodwater would have to be diverted into the Atchafalaya River and the West Atchafalaya Floodway, exceeding the design limits of these two outlets, not to mention hastening the Mississippi's channel jump. All of the Corps' observations indicted that this problem would only get worse.[15]

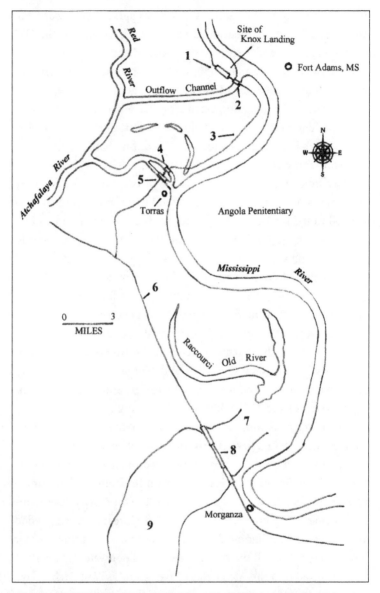

Figure 5.1 The Mississippi River channel between Old River and Morganza showing (1) overbank structure, (2) low sill, (3) Highway 15, (4) dam across Old River, (5) lock, (6) Highway 1, (7) Morganza Floodway intake, (8) Morganza control structure, and (9) Morganza Floodway.

The 1950 flood, one of the biggest on record, helped infuse the chief of engineers' task force with a sense of urgency.[16] Based upon the findings in *The Atchafalaya River Study*, the Mississippi River Commission asked the Corps to move ahead quickly with design studies and model experiments in preparation for a control structure and a navigation lock at Old River. In the brief time available, possibly less than ten years, the Corps had to deny the Mississippi its new path to the Gulf. The time frame would be less intimidating if the Corps could simply dam Old River to stop the impending channel jump, but navigation and flood-control interests rendered this option untenable.

The Mississippi's bifurcation could not be reversed; the only viable course of action was to freeze the divided channel at its present distribution, with approximately 30 percent escaping into the Atchafalaya. Although the appropriate structure might be able to maintain the 30/70 split, how to correctly divide the Mississippi's sediment load was a critical consideration. Too much mud remaining in the Mississippi below Old River would hasten the channel's deterioration above Morganza. There was a palpable feeling that the Corps would have only one chance to solve the sediment problem—they couldn't keep rebuilding the control structure until they hit upon the right design. It was this extraordinary set of circumstances that brought Hans Albert Einstein to Vicksburg in late 1951.[17]

The new task force charged with strategic planning for the Old River control structure and navigation lock included three sediment specialists. Besides Einstein, the Corps brought in Lorenz G. Straub, founding director of St. Anthony Falls Hydraulics Laboratory at the University of Minnesota and president of the International Association for Hydraulic Research, and one of its own, Donald Bondurant, Head of the Sediment Section with the Corps of Engineers' Missouri River Division. Einstein and Straub had worked with Bondurant for a number of years on the Corps' Missouri River projects. None of them had firsthand experience with the Lower Mississippi River's prodigious tonnage of clay, silt, and sand.[18]

For Einstein, the Old River problem was propitious. The previous year, the USDA Soil Conservation Service had published his landmark paper, *The Bed-Load Function for Sediment Transportation in Open Channel Flows*. In *The Bed-Load Function*, Einstein proposed dividing a stream's sediment load into two categories, wash load and bed load, to obtain a more accurate framework for understanding how a river transports soil. Wash load refers to smaller particles of sediment that travel suspended within the current, while bed load comprises the larger, heavier particles that move along the stream bottom by bouncing and tumbling.[19] When the Berkeley professor arrived in Vicksburg, the Corps' engineers were already busy crunching the Old River data with the equations from *The Bed-Load Function*. In the sediment-laden Lower Mississippi River, Einstein could hardly have chosen a more challenging and high-profile testing ground for his new method.

The consultants met four times with Corps and Mississippi River Commission engineers between December 1951 and May 1953. These gatherings in Vicksburg and New Orleans included excursions through Old River and down the Atchafalaya River on board the Corps' steamer, *Newton*.[20] Transcripts of these talks reveal the difficulty of the problem the Corps faced. Despite the urgent need to prevent the Mississippi River's diversion into the Atchafalaya Basin, an overriding concern during the meetings was the maintenance of the status quo at New Orleans. An Old River control structure must not raise river stages at the Crescent City or interfere with its navigation interests. And in spite of the capture threat, the Atchafalaya River's carrying capacity had to continue to increase in order to handle its share of the project flood. Any control structure in Old River must also keep an adequate flow going down Atchafalaya for sanitation, water supply, navigation, and irrigation. Finally, the project would need to stop the sediment accumulation in the Mississippi's channel below Old River.[21] As Lorenz G. Straub laconically remarked, the Corps' manifold dilemma was "something of a new problem."[22]

During the meetings, the Old River discussions exposed a rivalry between two differing approaches to predicting river sediment movement. The standard or traditional way to study sediment in a particular stream was straightforward: get into the water, take samples, and make predictions based upon empirical findings. Straub supported this methodology. Einstein, on the other hand, based his bed-load function method on laboratory testing with scale models to develop laws based upon the principles of mechanics. In part because it was new and its mathematical formulas conveyed a sense of scientific assurance, Einstein's approach inspired the Corps engineers. The bed-load function method held promise for predicting how the Mississippi's sediment load would react to the presence of one or more control structures in Old River, but problems with the mechanical approach surfaced during the December 1951 meeting.[23]

Straub pointed out large discrepancies between Einstein's formula results and actual sediment amounts measured in recent samples taken from the Atchafalaya River. Straub, who had also experimented with formulaic approaches to solving river sediment problems, felt that the dynamic interplay between currents, channel configuration, and soil particles in a living stream might be too complex to be described by a set of equations. Einstein was willing to concede that his figures needed adjusting, and he worked with the Corps engineers to make corrections to his formulas, but other problems with Einstein's method eventually led the army engineers to question its use at Old River.[24]

The backbone of Einstein's bed-load function approach was the separation of a river's sediment particles into the aforementioned wash load and bed load. On the face of it, hydraulic engineers should be able to distinguish between what is freely floating in the current and what is tumbling along the bottom, but the Mississippi was proving to be a difficult subject. When the group convened during 1952, disagreement on what constituted the two types of loads was a major stumbling block. Although it must have been hard for him, even Einstein admitted having difficulty drawing a clear line in the sediment.[25] The confusion is understandable; the Mississippi's river bottom is constantly changing. Sand waves as high as forty feet can undulate down the channel bed, and the bottom is hammered with rolling masses of waterlogged trees and other debris. This kind of interference, along with changes in discharge rate from one week to the next, can stir the bed load into suspension, giving it the characteristic of wash load.[26]

As a field test for Einstein's method, the Old River problem was disappointing. With time a critical factor, the Corps decided against using the formulas in favor of predictions based on the sediment data collected thus far coupled with ongoing tests using scale models. In fact, running controlled amounts of water and sediment through simulated river channels had been helping hydraulic engineers since the 1930s. The Corps' Waterways Experiment Station in Vicksburg operated Old River mockups of various sizes. In nearby Clinton, Mississippi, the engineers manipulated a working replica of almost the entire Mississippi River drainage system sprawled across nearly two hundred acres. Designed by the Corps and built during World War II with the help of German and Italian prisoners of war, the super model contained some fifteen thousand miles of waterways fit to a horizontal scale of 1:2000 and a vertical scale of 1:100.[27]

Beyond the debate over the use of Einstein's formulas, the most important questions considered by the Corps and its consultants centered on the control structures, which had to fit into the existing flood-control plan. Preliminary designs favored the use of two weirs placed side by side, one with enough depth to convey Mississippi water to the Atchafalaya River during times of low flows and another to help control water overtopping the river bank during high flows, respectively labeled "low sill" and "overbank." The group had to decide where to place the structures, how best to orient the intake channel in order to capture the proper amount of sediment, and consider structural design options. Everyone agreed that the structures should be located near the Mississippi River as opposed to alternate sites on the Atchafalaya River to better regulate the water going into the distributary. To the relief of New Orleans interests, Straub, Einstein, and Bondurant also felt that the Corps' dredging in the lower Atchafalaya, vital to the safe passage of big floods, could continue without affecting the operation of the control structures.[28]

Dredging in the Atchafalaya would, in fact, become even more important. The consultants recommended situating the Old River control structures to draw as much bed load material as possible out of the Mississippi in order to improve the channel capacity below Old River. This strategy would likely increase the amount of soil going into the Atchafalaya, making more work for the dredge boats. Capturing the bed load sediment depended upon the placement of the structures. The bed load moves in the slow part of the current in a meandering river. As discussed in Chapter 1, when the river rounds a bend, the fast current travels around the concave or out-side of the curve, while the current moves more slowly on the con-vex or inside of the bend. In addition to siting the low-sill structure's intake channel on the convex part of a bend, the angle of the intake relative to the main river channel has a bearing on the capture of bed load sediment. Placing the intake channel at a 90° angle or less will further slow the water coming in toward the structure and help convey more bed load material into the diversion. Of course, every-one realized that an important and unknown element in the plan was how the Mississippi might react to the intake channel. Rodney Latimer, co-editor of *The Atchafalaya River Study*, cautioned that the river could change its alignment relative to the intake or form a sandbar that would interfere with the control structure.[29]

After considering alternate locations, the team selected a site known as Knox Landing, opposite the Mississippi town of Fort Adams and about five miles north of Old River (see Figure 5.1).[30] Although additional borings were needed to complete the site's geo-logical assessment, the Corps' records showed that the river's bank line here had remained stable since the late nineteenth century. Knox Landing is also situated in a convex bend position where the

Mississippi flows around Point Breeze, with plenty of the bed load sediment moving along the west side of the Mississippi. The location looked so favorable that Straub even expressed some concern that the intake could capture too much bed load material and clog the outflow channel that would carry the water from the structure to the Red/Atchafalaya. Continued experiments with scale models helped the Corps analyze this and other potential problems.[31]

Meanwhile, the task group considered and discarded several different designs for the control structures and vetoed a bizarre suggestion by Straub to dispense with a structure and lock in favor of an open Old River channel lined with derrick stone and riprap. In a long statement that must have raised some eyebrows around the table, Straub maintained that piling rocks in Old River would provide sufficient "friction control" to regulate the flow of water and sediment into the Atchafalaya. Despite its low-cost appeal, Straub's radical idea was, of course, unfeasible. Old River needed formidable concrete and steel structures built to give the Corps secure control over a desperate situation, with the flexibility to handle the Mississippi's annual rise and fall. The low-sill structure's design evolved to try and meet those needs.[32]

So named because of the depth of the threshold or sill at the bottom of its eleven steel gates, the low-sill's three center gates sit five feet below mean sea level and the remaining eight gates—four on each side—are ten feet above mean sea level. The structure's depth provided the necessary flexibility to permit operation during most low water seasons. Cost considerations precluded a lower sill design and the river's fluctuating depth, ranging from ten to one hundred feet below mean sea level, meant that a portion of the desirable bed-load sediment would probably miss the intake channel. The overbank structure, with its lower gate sill at fifty-two feet above mean sea level, was being designed to work in tandem with the low sill and, as its name suggests, handle high flows that overtop the riverbank. Like the Bonnet Carré structure, the overbank structure would operate with needle-type gates. Important details remained to be worked out, such as the best angle of diversion into the intake channel and the elevation of the stilling basin behind the structures to prevent scouring in the outflow channel.[33]

While Waterways Experiment Station engineers ran tests with scale models to confront these issues, the Corps and consultants worked on the plan for the project's other component, the navigation lock. The lock site incorporated part of Old River's lower arm, about seven miles south of the low-sill and overbank structures (see Figure 5.1). Although this was vital for keeping river traffic moving through Old River, the consultants were also interested in the lock's potential to help the low sill keep water flowing down the Atchafalaya during times of low water on the Mississippi.[34]

While completing the basic plans for the lock and control structures, the Corps and the Mississippi River Commission considered themselves in a race against nature. Brigadier General John R. Hardin, president of the Mississippi River Commission, put it succinctly in an article he published in the journal *The Military Engineer* in early 1954: "It is imperative that construction be initiated promptly and proceed in a prescribed order, as the earliest date when all construction can be completed under favorable conditions very nearly coincides with the date when it is estimated that the earliest possible critical stage of development in [the] capture of the Mississippi by the Atchafalaya will be reached."[35]

◆ ◆ ◆

BY "THE EARLIEST POSSIBLE CRITICAL STAGE," General Hardin was referring to the 1960 point of no return predicted in Latimer and Schweizer's *The Atchafalaya River Study*, when the Mississippi's diversion into the Atchafalaya would be too far advanced to stop. His "prescribed order" of construction placed the low sill first, followed by the overbank structure, with the lock third in line. Once the lock was completed, the Corps had to dam Old River's lower arm. Hardin, who took over as head of the Mississippi River Commission and the post of Lower Mississippi Valley Division Engineer in 1953, wasted no time in seeking congressional authorization for the battery of new structures. Armed with *The Atchafalaya Study*, army engineers provided lawmakers with ample justification for the project. The Mississippi River Commission estimated that the Atchafalaya's capture of the Mississippi River could result in losses ranging from $400 million to just over $1 billion. Members of Congress also heard from anxious private interests such as the Atchafalaya Basin's oil and gas industry, sugar producers, and the railroads that crisscrossed the basin.[36]

Congressional confidence in the Corps of Engineers helped to speed the inclusion of the Old River control project in the Flood Control Act of 1954, with authorization to spend $47.4 million. Although the bill's language specified the navigation lock as part of the project, funding for this component, approximately $28 million, was not authorized until 1958. To guide the final planning into the construction phase, Hardin formed a new task group called the Board of Consulting Engineers for Old River Control, with Einstein, Straub, and Fisk continuing in their advisory roles. By the time of the group's first meeting in early 1954, engineers at the Waterways Experiment Station were ready with model demonstrations of the low-sill control structure and bed-load simulation.[37]

Geological borings at the Knox Landing site where the low sill would be constructed confirmed the presence of deep sand deposits, remnants of the ancient river channels described in Chapter 1. The Corps would have preferred a better geological situation, but circumstances limited the placement of the low sill to this critical reach along the Mississippi's west bank. After all, Knox Landing was well-suited for capturing as much bed-load sediment as possible. In the Corps' favor, the borings revealed that the low sill and part of its outflow channel could be set in a more recently abandoned riverbed running perpendicular to the Mississippi's main channel. This old channel formed a trench about 120 feet deep and about 1,000 feet wide filled with silt and mixed silt and sand that cut through

the surrounding river sand left by more ancient Mississippi channels. The trench fill would facilitate excavation for the structure and
provide a more stable foundation than the older sand deposits. The
geological borings also showed that the outflow channel behind the
low-sill and overbank structures would traverse a patchwork quilt
of deposits that included areas of channel fill related to abandoned
riverbeds similar to the one below the low sill as well as clay backswamp deposits and old point bar deposits composed of silts and
sands, which all overlie the deeper sands. Corps engineers felt that
the dense clay deposits would help stabilize the outflow channel
during long-term use.[38]

◆ ◆ ◆

DIRT BEGAN TO FLY at Knox Landing in September 1955, and the
Corps completed the low-sill control structure with its intake and
outflow channels in June 1959, just six months shy of the 1960 critical threshold forecast in *The Atchafalaya River Study*.[39] The structure tasked to permanently tame the Lower Mississippi River is a
barricade of steel-reinforced concrete approximately 75 feet high
and 566 feet in length (see Figure 5.2). A two-lane highway crossing
the top of the low sill runs alongside the tracking for a one-hundred-ton rolling gantry crane for raising and lowering the eleven
44-foot-wide gates.[40] Anchoring the structure to the floodplain
are vertical and batter (angled) steel H-beam pilings driven to a
depth of approximately 90 feet through the silt layer into the top of
the underlying sand formation. On the Mississippi River side, the
Corps designed the low sill to withstand high water up to 69.8 feet
above mean sea level, which is some 17 feet above the riverbank.[41]

Although a forest of deep-set pilings secured the low-sill's main
edifice to the earth beneath it, the structure's curving, concrete wing
walls (also called training walls) designed to channel the Mississippi's water into the gates were not fortified with pilings. Sixty-seven
feet high, attached at one end to the low-sill face, and standing on

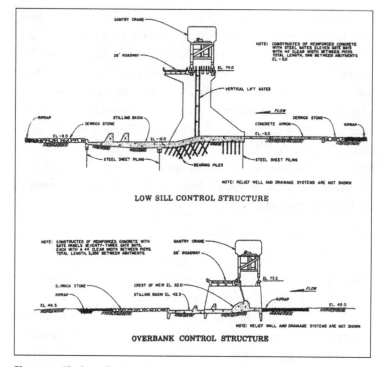

Figure 5.2 The low-sill and overbank control structures shown in cross section. Fairless, "The Old River Control Project," Figure 3. Image courtesy of the American Water Resources Association.

the intake channel's concrete apron, the wing walls extend 140 feet out into the fore bay (see Figure 5.3). In light of what would happen during the 1973 flood, the omission of wing wall pilings was a serious design error. Cost-saving measures influenced other aspects of the low-sill's construction; however, the decision to go without pilings in this critical part of the structure seems to have been justified by the finding of a Corps study of 1958 reporting that "computations show that the [wing] walls should have adequate stability without the necessity for a pile foundation."[42]

Work began on the low-sill's companion weir, the overbank structure, in October 1956 (see Figure 5.2). Using the Waterways

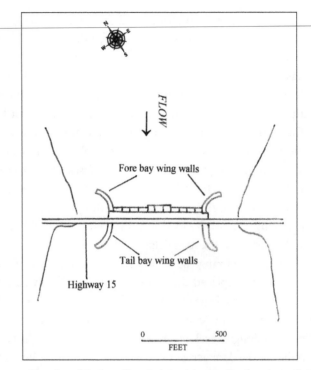

FLOW

Fore bay wing walls

Tail bay wing walls

Highway 15

0   500

FEET

Figure 5.3 Plan view of the low-sill control structure showing the wing walls in the fore bay (Mississippi River side) and stilling basin. Based upon Moore, "Structures Required," Figure 2.

Experiment Station's scale models, the Corps experimented with options for placement of the structure, eventually settling on the site adjacent to the north end of the low sill. The structure is 3,356 feet long with seventy-three bays, each bay holding fifteen gates. Each gate comprises three eighteen-foot-long needle logs, each 11.5 inches wide and 10 inches thick, bolted together side by side. The same two-lane highway crossing the low sill spans the overbank structure, which has two rolling gantry cranes to operate the gates. Unlike the low sill, the overbank is not anchored by pilings. When the Corps completed work on the overbank weir in October 1959, the paired structures stood ready to repel the Mississippi's channel jump. General Hardin estimated that together the low sill and overbank could pass up to seven hundred thousand cubic feet per second.[43]

The last items on the Corps' Old River checklist were the navigation lock and final damming of the lower arm of Turnbull Bend. Excavation commenced in July 1958 for the lock's 1,185-foot-long chute. The original plans specified a single chamber 84 feet wide and 80 feet deep, with the lock's floor set at 11 feet below mean sea level. Perhaps as another measure to lower project costs, the lock chamber's width was reduced to 75 feet by the time construction began.[44] Completed in December 1962, the lock began operating in March of the following year. At the lock site, the average head differential between the surface of the Atchafalaya and the higher Mississippi is about 16 feet and can range up to around 19 feet during big floods. The new structure replaced the east end of lower Old River, while traffic continues to navigate the familiar dogleg passage between the lock and the head of the Atchafalaya River (see Figure 5.1).[45]

Following a design developed by Hans Albert Einstein, the Corps began building the dam across Old River's lower arm in June 1963. The dam site is immediately north of the lock and approximately 860 feet west of the abandoned railroad bridge over Old River where the river town of Torras once stood (see Figure 5.1). Einstein recommended doing the work after the Mississippi's annual spring rise had crested and begun to fall. To prepare the closure, the Corps laid a mattress of interwoven boards across the bottom of the channel and began dumping riprap stone and gravel to form the base of the dam. Eventually, the stone structure's crest reached the target height of 17 feet above mean sea level, at first submerged but later exposed as the Mississippi dropped through the low-water season. For the dam's next stage, two cutterhead dredges sucked up Old River's sand and redeposited it over the stone structure, raising the crest to 45 feet above mean sea level. To complete the dam, the Corps brought in enough clay to encase the structure and raise the crest another 23 feet, winding up the project in only four months.[46]

With the Old River closure, the Corps assumed control over the three rivers junction. If the hard-won research proved true, the new structures could maintain the Mississippi's 30/70 split at practically all river levels. For the Corps of Engineers and the Mississippi River Commission, the timely completion of the Old River control structures attests to the strength of their emergency response partnership with Congress. Writing for *The Military Engineer* in 1964, Corps Colonel Edward B. Jennings stated confidently that "a threat of catastrophic proportions has been ended forever." Meanwhile, the Mississippi began to adjust to yet another human modification at Turnbull Bend. Just a few years later, the 1973 flood would challenge the Corps' use of words like *forever* when discussing the success of its mission at Old River.[47]

CHAPTER 6

The Real Crisis

On April 11, 1973, Lieutenant General Frederick J. Clarke, chief of engineers, was on Capitol Hill seeking to reassure members of the House Committee on Public Works. His briefing offered the representatives a fleeting distraction from the Watergate controversy already swollen beyond White House control. Clarke was happy to report that the Corps of Engineers was holding in check one of the largest Mississippi River floods in history.[1]

General Clarke: "I think we can say that the real crisis is almost past."

House Committee Chairman, John A. Blatnik of Minnesota: "Would you repeat that? The real crisis is almost past?"

Clarke: "Is almost past."[2]

Three days later, the rampaging Mississippi swept away part of the low-sill control structure at Old River. The real crisis was about to begin.

◆ ◆ ◆

BEFORE THE GREAT FLOOD OF 1973, the Corps of Engineers' confidence in the Old River control structure complex eerily recalls a similar level of confidence in the disastrous levees-only strategy on the eve of the 1927 flood. In 1958, Norman R. Moore, chief of the Mississippi River Commission's Engineering Division, wrote that the Old River structures "will be a safe and permanent solution to the diversion problem."[3] As mentioned in the last chapter,

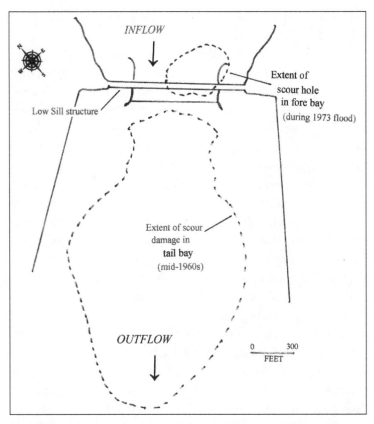

Figure 6.1 Plan view of the low-sill control structure showing the extent of the mid-1960s scour damage in the tail bay and the extent of the scour hole in the fore bay caused by the 1973 collapse of the south wing wall. Based upon Fairless, "Old River Control Project," Figure 4, and Hebert, "The Flood Control Capabilities of the Atchafalaya Basin Floodway," Illustration 8.

Colonel Edward Jennings, New Orleans District engineer, declared everlasting victory over the Mississippi's impending avulsion.[4] The 1964 annual report of the chief of engineers put it in matter-of-fact terms: "The Old River control project will prevent the steadily enlarging channels of the Old and Atchafalaya rivers from capturing the flow of the Mississippi River."[5] But as soon as the control structures went into operation, the river began to knead and gnaw at the artificial outlet.

The concern that the Mississippi's relentless scouring might undermine the low sill surfaced during the Corps' meetings with Einstein, Straub, and Bondurant in the early 1950s. Rodney Latimer, co-author of *The Atchafalaya Study*, was particularly worried about the possibility of damage to the stilling basin on the outflow side of the low sill.[6] Here the Corps' artificial channel would have to endure the steady, grinding force of sediment-laden water pouring through the structure's gates at a rate of hundreds of thousands of cubic feet per second. To cope with this continuous impact, the engineers designed the stilling basin with a steel-reinforced concrete floor studded with raised concrete baffle blocks to slow and diffuse the current.[7] Yet despite the design's appearance of durability, the Mississippi River soon justified Latimer's fears.

Only a few months after the low sill began operating, Corps personnel taking soundings in the tail bay discovered a hole scoured to a depth of around 30 feet below the floor of the outflow channel. Despite the dumping of barge-loads of riprap stone, the Mississippi doggedly chewed at the weak spot until the hole became a crater. By 1964, the hole was 1,200 feet wide and an alarming 130 feet deep, almost forty feet lower than the deepest pilings anchoring the low sill (see Figure 6.1). Fortunately, the scouring was progressing westward down the outflow channel away from the structure. After dropping another 150,000 tons of stone into the crater, the Corps remained uncertain about the permanency of the repair job.[8]

While the engineers tried to cope with the damage to the tail bay, another serious problem developed on the inflow side of the low sill. In 1964 and 1965, loose barges sucked into the structure's inflow channel from tow mishaps on the Mississippi River collided with the low-sill's gates. Some of the barges loaded with tons of riprap became wedged against the gate bays and compromised the structure's operation. Closing the low-sill's gates to extract the barges and repair the damage took several months. With little or no water passing through the structure during this time, the increased head differential—the rising water level in front of the structure compared to the low level on the outflow side—stressed the low-sill's design limits. More serious, however, was the potential for additional scouring in the tail bay when the gates finally reopened.[9]

During the 1970s, hydraulic engineers at the Corps' Waterways Experiment Station in Vicksburg tested various schemes to deflect runaway barges. In the meantime, a picket boat patrolled the low-sill's inflow channel during times of high water to protect the structure's gates.[10] Besides the potential structural problems posed by the barge collisions, closing the low sill for repairs interrupted the maintenance of the 70/30 percentage split between the amount of water remaining in the Mississippi's channel (70%) and the amount going down the Atchafalaya River (30%). Since the barge collisions in 1964 and 1965 happened during the spring rise, the low-sill's temporary closure for repairs brought unexpected relief to the farmers of the Red River Backwater. These welcome reprieves from the annual backwater flooding prompted requests to close the low sill every year during the planting season.[11]

Besides the danger of increasing the structure's head differential during the annual high-flow season, New Orleans interests would have probably opposed the temporary loss of the city's upriver safety valve. Likewise, the people depending upon the Atchafalaya River would have raised loud objections to any annual reduction in their share of the Mississippi. In addition to the adverse effects on the Atchafalaya Basin's agriculture, transportation, and recreation, an extended reduction in the Atchafalaya River's flow could allow the intrusion of Gulf saltwater, threatening Morgan City, Berwick, and the other towns near the Wax Lake and Atchafalaya River outlets.[12]

But these concerns were gradually becoming overshadowed by a deeper fear. Before the construction of the Old River control structures, engineering studies found that the lower Atchafalaya Floodway, below the merger of the Atchafalaya River with the Morganza and West floodways, was not capable of conveying half of the project flood, 1.5 million cubic feet per second, down to the Gulf. The problem was the steady accumulation of sediment clogging this crucial part of the floodway and the tendency of the floodway's guide levees to sink into the swamp under their own weight. Follow-up studies done in the late 1960s revealed that the problem had gotten worse since the installation of the structures at Old River.[13] As the decade of the 1970s dawned, a viable solution to the project flood dilemma still eluded the Corps.

◆ ◆ ◆

HARD RAINS BEGAN in October 1972, much heavier than usual for that time of year, and by January 1973 the weather pattern had become uncomfortably similar to records for the major flood years of 1927 and 1937. Other ominous signs appeared as well. At Cairo, Illinois, Mississippi River gage readings during October and November 1972 were five to ten feet higher than those recorded for the same months in 1927 and 1937 as hundreds of thousands of acres of saturated ground passed the runoff from ditch to creek to stream to river. From St. Louis to New Orleans, December brought record high river stages. Up the valleys of the Red, Arkansas, White, and Yazoo Rivers, the great reservoirs constructed after 1927 were quickly filling to capacity.[14]

In November 1972, Colonel Richard L. Hunt, New Orleans District engineer, briefed the Association of Louisiana Levee Boards meeting in New Orleans. By coincidence, 1973 was a year for the Corps' biannual flood fight exercises with state, local, and federal agencies. During February, while the real flood gathered strength, teams mobilized throughout the Mississippi Valley to fight a mock deluge, which included testing the gate bays at the Bonnet Carré, Morganza, and Old River structures. That same month, Major General Charles C. Noble, president of Mississippi River Commission and Lower Mississippi Valley division engineer, testified before Congress. Noble, having held his post since 1971, recognized the need for additional manpower, equipment, and supplies to raise many of the existing levees. Because of the ever-changing riverbed, levee heights would have to be increased in places to maintain the preferred three-foot free-board or margin between the crest of the river and the top of the levee.[15]

In early March, more storms brought prodigious rainfall across the Missouri River Basin and the Upper Mississippi. By March 9, the Corps knew that a major flood, possibly one on the scale of the dreaded project flood, was brewing. Compounding the difficulty, the Mississippi Valley reservoirs were now full and water needed to be released to take pressure off of the dams.[16] As the rising swell moved down the Mississippi Valley, formal warnings went out to governors, levee boards, and other state and local officials. The Corps also notified property owners to remove all livestock grazing the levees and open access gates for flood fighters. Fortunately, the mock flood fight conducted just weeks earlier saved time in inspecting levees and control structures. Anticipating the worst, teams at flood fight command centers began assembling volunteers and assessing the need for resources to equip workers.[17]

◆ ◆ ◆

LIKE THE 1927 FLOOD, the flood of 1973 crested several times. The first crest came early, peaking in December and January before the Mississippi began dropping through February and leveling off in March. By March 19, increasing river stages signaled a second rise on the upper river fueled by the Missouri Basin storms. In early April, the third and highest crest moved past St. Louis and took aim at the Lower Mississippi Valley.[18]

Far to the south, residents of the Red River Backwater region faced their first major flood since the installation of the Morganza Floodway and the new control structures at Old River. Many people along the Red, Black, and Tensas Rivers believed these structures would decrease the backwater buildup, but as the floodwaters descending both the Red and Mississippi rose higher, they soon found their situation worse than before. On April 1, Red River flooding forced the Corps to begin the evacuation of Avoyelles Parish residents in the Big Bend area along Bayou des Glaises.[19] The next day, families in Jonesville, Catahoula Parish, had to evacuate to escape the rising Black River.[20]

By late March, more than two hundred thousand acres in Concordia Parish were under water. Many parish residents, including State Senator James H. Brown of Ferriday, thought the Corps of Engineers could relieve the situation. Apparently misinformed about which floodways affected his district, the senator initially wanted the Corps to open the Bonnet Carré spillway, but the backwater parishes soon shifted their hopes to Morganza and Old River. Some residents believed that the Corps could reverse the rising water by closing the Old River structures and opening the Morganza Floodway. With the third crest bearing down on the Lower Valley, desperation quickly led to anger. On April 3, State Representative Raymond La Borde of Marksville reported receiving word of threats to dynamite the Morganza structure. In response, Corps personnel at Morganza went on double duty, working night shifts to patrol the spillway. A series of well-informed articles and editorials in the *Concordia Sentinel* newspaper helped to diffuse the tension by allowing Corps officers to explain why the low-sill control structure was built and why it could not be closed. Corps engineers also informed the backwater residents that opening the Morganza Floodway would likely increase their flooding.[21]

On April 1, the Corps had to revise upward the crest predictions for the Mississippi at Old River and points south following recent rains in the Lower Valley. Preparations for use of the Bonnet Carré and Morganza structures were well underway while the Corps awaited the river conditions that would require the activation of these spillways. Not surprisingly, the engineers faced pressure from opposing interest groups advocating the use of one floodway before the other. New Orleans interests wanted a Morganza opening in order to tame the flood as far upriver as possible. Lake Pontchartrain fishing interests also favored using Morganza first, fearing the impact that the Bonnet Carré spillway could have on the lake's fish and shellfish. Naturally, people in the Atchafalaya Basin wanted Bonnet Carré opened first.[22]

The New Orleans gage hit 19 feet early on April 8 with a National Weather Service prediction that the swell would rise to 20.2 feet later in the day. According to the National Oceanic and Atmospheric Administration, the New Orleans levees are designed to protect the city at river stages up to 20 feet on the local gage. For the Bonnet Carré spillway, the waiting was over. From his office at Vicksburg, General Noble telephoned the decision to open the structure for the first time since 1950. Expecting protests, Corps personnel were pleasantly surprised when more than two thousand people gathered along the levees lining the spillway to watch Louisiana's US senator Russell B. Long take the controls of the gantry crane and pull the first needle logs out of their gate slots. The Corps' official account of the 1973 flood called the Bonnet Carré opening "a happy, almost gala occasion." However, if the crowd expected to see the dry floodway suddenly filled with a mighty torrent, they were disappointed; the Corps prudently opened the Bonnet Carré gates slowly to avoid scouring, stretching the operation over four days before water poured through all 350 gate bays.[23]

The next day, the *Baton Rouge Morning Advocate* carried the provocative headline "River Water to Take Toll In Seafood," with an accompanying article lamenting the sacrifice of Lake Pontchartrain's oyster, crab, and shrimp industry in order to keep the Mississippi out of New Orleans. But environmentalists countered that reports of the demise of the lake's fish populations were premature. Instead of killing the aquatic life in Lake Pontchartrain, the infusion of Mississippi River water would supply the lake with helpful nutrients. Indeed, in the Corps' 1973 flood report, General Noble noted that seafood industry protestors "made more noise before [the Bonnet Carré opening] than after."[24]

The Bonnet Carré spillway opening kicked off what would become a desperate flood fight at the lower end of the Mississippi Valley. Generals Clarke and Noble didn't know it yet, but the real crisis was riding the flood's third crest. Two days after the activation of the Bonnet Carré spillway, the river level at Carrollton had only dropped about six inches and anxious officials in New Orleans began to look north toward Morganza. The great levee crevasses that saved the city in 1927 wouldn't recur this time thanks to the Corps' formidable mainline levee system. Now the Crescent City's well-being depended upon the Atchafalaya Floodway.[25]

During the second week of April, the low sill was sending around 500,000 cubic feet per second down the Atchafalaya River. The overbank structure remained closed but could pass about 375,000 cubic feet per second if necessary. The Morganza Floodway would need to nearly match the low sill and overbank discharge in order to draw half of the flood out of the main channel and keep the flow past New Orleans at a safe level. If necessary, the West Atchafalaya Floodway could probably pass another 200,000 cubic feet per second. Following established procedure, the Corps alerted people who had livestock and property in the Morganza Floodway. General Noble also issued orders to raise the Morgan City floodwall and announced that he would give five days' notice before opening the Morganza gates.[26]

As it turned out, the Morganza Floodway would have to be activated with hardly any warning at all.

◆ ◆ ◆

LATE SATURDAY AFTERNOON, APRIL 14, two engineers from the Corps' New Orleans District headquarters, Tom Johnson and Richard Hill, left the Old River Control Complex for the thirty-one-mile drive back down the levee road to the Morganza Floodway. This had been another long day with no real cause for concern. While the flood fight raged, the pair commuted daily from their temporary lodging at Port Allen near Baton Rouge to inspect the two flood control sites. Under normal river conditions, resident foremen managed the operations at the Old River and Morganza structures, but because of the flood situation, Johnson and Hill had assumed temporary authority over the two installations.[27]

Forrest Bordelon had been the Old River Control foreman since the low-sill and overbank structures began operating.[28] Johnson and Hill were gone less than an hour when Bordelon walked out on the low sill to check the structure before supper. Eight-hundred yards to the east, the Mississippi River was in full flood, throwing one-fourth of its discharge down the intake channel toward the low-sill's gates. Beyond a steady vibration, the massive concrete edifice seemed to be holding up to the punishment. As Bordelon ran his eye over the structure in the fading afternoon light, he realized that a frightening change had occurred in the fore bay—part of the wing wall at the south end of the intake bays was gone. In place of the wing wall section, a swirling eddy pummeled the remaining length of wall attached to the face of the structure. Moments later, Bordelon sat down at his office radio to call for help.[29]

From the perspective of a twenty-first-century world bristling with cell towers and satellite transmissions, it is hard to imagine that the Old River control complex did not have telephone service in 1973. But aside from Forrest and his wife Frances, who occupied the foreman's residence beside the low-sill structure, no one lived within miles of Old River and telephone lines had not yet been strung down Highway 15 along the west side of the Mississippi. As a consequence, Forrest Bordelon's only connection with other Corps of Engineers personnel was the installation's two-way radio. He could transmit to his staff's vehicle radios and to the 210-foot tower down at the Morganza Floodway control structure, although the people he needed to talk to, Johnson and Hill, happened to be traveling in a government car without a radio set.[30]

About 6:30 pm Perry Gustin, assistant structure foreman, was catching up on paperwork in the Morganza control structure office when his radio crackled. The next thing he heard was Bordelon's agitated voice. Having worked at the Old River complex, Gustin was quite familiar with the low-sill's wing walls. He listened to the dire news and asked, "Mr. Forrest, you mean the curved part? That fell in? You don't see it?"

Bordelon answered, "Kid, it's gone."[31]

When the two engineers finally reached the Morganza office, Gustin quickly put Johnson in radio contact with the Old River foreman. The update was grim. In the brief interim since Bordelon's initial call, a second section of the wing wall fell, leaving only a short fragment standing.[32]

The Morganza office had a duplicate set of drawings and specifications for the low-sill structure and a telephone. While Gustin spread out the engineering plans, Johnson telephoned his supervisor, Frederick Chatry, chief of the engineering division, New Orleans District. Chatry took the call at his home and, after a brief exchange, decided that the low-sill's gates should be closed. Although undocumented, Chatry may have feared that the water rushing through the structure was enlarging the scour hole in the tail bay. If Tom Johnson wasn't already aware of the weakness in the low-sill's wing wall design, his examination of the structure's plans and specifications would have revealed the lack of anchor pilings.[33]

While Johnson studied the drawings, Perry Gustin telephoned for additional help. They needed more hands to close the low-sill's gates, and Gustin was able to rouse the two-man crew that normally worked with Bordelon at Old River. With reinforcements on the way, Gustin drove the Morganza truck up the darkening highway to Old River followed by the two engineers. In those days, few vehicles traveled the levee road to the control structures, especially at night.

Forty-five minutes later, an anxious group of men, including Bordelon, Johnson, Hill, Gustin, and Leroy Dugas, Morganza Floodway foreman, gathered in the dark at the south end of the low sill. Someone aimed a flashlight beam out across the flooded intake channel and found only churning froth where the south wing wall once stood, but the low sill held strong. The amount of water rushing through the structure that night was significantly higher than the low-sill's design limit of around 325,000 cubic feet per second. This flood was the structure's first real test. Nobody knew if the hydraulic calculations and scale model experiments could approximate a real Mississippi River flood, with its bullwhip current choking with sediment and drifting debris. And the engineers and foremen standing beside the dark tempest in the intake channel could not know what changes were taking place down on the fore bay floor at the low-sill's mouth. All that was really apparent was that the structure's eleven gates were handling the load. Now that he was on the scene, Johnson had second thoughts about what might happen if they began closing those gates.[34]

On paper, the low sill could tolerate a head differential up to thirty-seven feet. Closing the gates would quickly pile up the water in the fore bay and drop the level in the tail bay. Under ideal circumstances the structure should be able to withstand the strain, but the Corps was fighting a monster flood, and the low sill had already sustained an unknown amount of damage. Indeed, after the flood, Corps engineers determined that the structure could probably only withstand an eighteen-foot head in its weakened condition. Recalling the gate question several years later, General Noble reflected that he ordered model testing during the flood to determine whether the bays should be opened or closed. But on that wild Saturday night at Old River, Tom Johnson was the senior Corps of Engineers official in charge, and he had his orders. The Mississippi River wasn't going to allow him the luxury of time for tests and computations. What happened next helped to avert a catastrophe. Johnson told Bordelon to go ahead and prepare the low-sill's crane for gate closure while he returned to Morganza and the telephone to share his firsthand observations and misgivings with Fred Chatry.[35]

With Johnson driving off into the night, Bordelon and Gustin could feel the control structure's vibration growing more violent. At the north end of the low sill, the one-hundred-ton, fifty-foot-high

Moffett gantry crane loomed in the darkness.[36] Because of possible damage at the site of the fallen wing wall, the south end gates would have to be closed first. Perry Gustin recalled the ensuing discussion: "Forrest turned to me, he said, 'Tell you what kid, I'm not riding that crane across that structure.' He said, 'I'll start her up on the other end and you catch her when she comes down on this end.' I said, 'OK, Captain I'll catch her.'"[37]

While the idea of a driverless crane rolling across the top of the low sill might sound reckless, Perry Gustin noted that the huge piece of equipment tracked quite slowly. In high gear, the crane's trip from one end of the structure to the other took thirty or forty minutes.[38]

If Johnson came back and his orders were unchanged, the gate-closing process would also be quite slow. Each gate is made up of stacked leaves or sections with each steel leaf measuring twenty feet by forty-four feet and weighing between twenty-five and thirty tons. The three center bay gates, being taller owing to their added depth, comprise four leaves each, while the eight outer gates each have three leaves. In opening and closing the structure, the crane must lift and position one leaf at a time. When the gantry finally reached the south end, Gustin climbed aboard and moved it off the structure onto a maintenance track extension to await the decision on closing the gates.[39]

Tom Johnson had departed about 9:00 pm and he returned from Morganza around midnight. It is probable that General Noble had been alerted in the interim and that Fred Chatry conferred with the general. At any rate, Johnson brought the news that the orders had been reversed. The gates were to remain open.[40]

◆ ◆ ◆

PALM SUNDAY, APRIL 15, found the low sill damaged but still standing, but for how long no one knew. The pressure was inexorable; one-fourth of the flood continued to shake the stricken structure. Although a few Corps engineers and other government officials were aware of the previous night's drama, it would be another day before the story became widely known. Instead, the Sunday editions of the Baton Rouge and New Orleans newspapers gave readers cause for optimism about the flood fight. A *Times-Picayune* article by Vincent Lee reported that opening the Morganza Floodway was "still not foreseen," and an editorial by Howard Jacobs in the same issue noted that "little necessity was seen for putting [the Morganza Floodway] plan into effect."[41] But by the time those newspapers hit readers' doorsteps, the Mississippi River had forced a change in plans.

At Old River, Tom Johnson was no doubt relieved to have someone of higher rank finally on the scene. That the individual on the ground and in charge was Major General Charles C. Noble indicates just how critical the low-sill dilemma was. Although the general immediately ordered the opening of all of the overbank structure's seventy-three gate bays, Noble realized that he would also have to quickly open the Morganza Floodway if the low sill was to be saved. Beyond salvaging the control structure, General Noble was considering the infinitely higher stakes at risk if the Mississippi swept away the low sill and the mighty river changed its course.[42]

The story broke on Monday, April 16. From New Orleans to Baton Rouge to Jonesville to Morgan City, people read about what happened at Old River and learned about the Corps' plan to open the Morganza Floodway in twenty-four hours. Louisiana governor Edwin Edwards quickly found himself between two opposing constituencies. New Orleans interests were quite happy with the decision, while protests rained on the governor from the Atchafalaya Basin and Morgan City. In desperation, Governor Edwards telephoned General Noble and asked, "Do I have authority to tell you not to open [the Morganza] floodway?" The general answered, "No."[43]

The telephones at the Corps' Vicksburg and New Orleans offices were also ringing. Farmers preparing fields throughout the floodway were going to be denied their spring planting. The loudest objections came from Morgan City and Berwick, where Corps personnel were already fortifying the levee tops with interlocking walls of steel sheet piling. General Noble recalled that Morgan City mayor Dr. Charles Brownell Jr. accused the Corps of "sacrificing Morgan City to save New Orleans."[44] Noble also noted that the mayor and other Louisiana politicians were fully aware of what could happen to the lower Atchafalaya Basin should the Mississippi River overwhelm the low sill. At Morganza, Corps officials hurriedly alerted people with property still remaining in the floodway.[45] When reminded of his promise of five days' advance warning, Noble countered, "The river didn't give us five days [notice]."[46]

The next day, Tuesday, April 17, Governor Edwards stood on the Morganza structure dressed in a gleaming white jumpsuit and black patent leather boots. If he was unable to prevent the activation of the floodway, at least he knew how to take advantage of a brief national spotlight. Not to be outdone by Senator Long's bravura performance at the Bonnet Carré spillway, Edwards climbed behind the controls of one of Morganza's two twenty-five-ton cranes to raise the first steel gate, activating the control structure for the first time since its completion twenty years earlier. Saving political face, the governor announced to the press that he and the Corps of Engineers had reached a compromise: only one-third of Morganza's gates would be opened to lessen the floodway's impact on the people of the lower Atchafalaya Basin. The Corps also granted floodway wildlife a brief reprieve and closed the Morganza gates for a while later that day to allow the Louisiana National Guard, volunteers, and Corps personnel to move deer, bears, turkeys, and other animals out of areas to be flooded.[47]

The Morganza opening brought a little relief to the people in the Red River Backwater by lowering the Mississippi's height at Old River. As many had feared, the emergency opening of the overbank structure on Palm Sunday had sent more water into the flooded Red, Black, and Tensas rivers. To make matters worse, the region received three days of rainfall totaling around eight inches.

Lingering local questions about the purpose of the Old River control structures prompted *The Concordia Sentinel* to allow General Noble the opportunity to once again explain the Corps' mission to prevent the Mississippi River from changing its course.[48] The general shared the Corps' dilemma, saying, "For every request I get to open [the control structures], I get just as many not to open them"[49] In early May, over 80 percent of Catahoula Parish was still under water despite the Corps' continuous efforts to raise the Black River levees with sandbags. Bulldozers also bolstered the mainline Mississippi levee near Old River, where the Corps feared a crevasse could bypass the control structures. Over in the West Atchafalaya Floodway, residents nervously watched the rise of Bayou des Glaises on the other side of their fuseplug levee.[50]

Thanks to the Morganza Floodway, the Mississippi's pressure on the low sill abated somewhat. A relieved General Noble marshalled cranes and barge loads of riprap to patch the old wound in the damaged tail bay and the new hole beneath the lost wing wall. Over the days and weeks that followed, a new wing wall made of boulders rose out of the low-sill's fore bay, one that dwarfed the graceful concrete structure it replaced. Curving out from the intake gates like its predecessor, the mountainous dike looked as tough as the structure it served.[51] With the flood subsiding by the fourth week in May, Corps engineers could afford to cautiously pat each other on the back. A major disaster had been narrowly averted. With the danger flowing into the past, Colonel Richard Hunt, the engineer directly responsible for the Old River Control structures, could not suppress a touch of hubris when he told Sam Hanna, editor of *The Concordia Sentinel*, that he "never was afraid of losing the [low sill]."[52] What the colonel and the rest of the rest of the Lower Mississippi Valley didn't know at the time was the extent of the damage the river had inflicted on the Old River structure. The flood's full impact would not be discovered until 1986, when a new control structure at Old River allowed the Corps to drain the low sill's inflow and outflow channels for a thorough inspection.

◆ ◆ ◆

"TOO CLOSE FOR COMFORT" is how General Charles Noble compared the 1973 flood to the Corps' three-million-cubic-feet-per-second project flood.[53] In his 1981 oral history interview conducted by Corps historian Martin Reuss, Noble added that the 1973 event "could have easily [reached the size of the project flood] with just a small difference in the weather pattern upstream."[54]

Although the estimates vary on the cubic-feet-per-second volume, the National Oceanic and Atmospheric Administration (NOAA) considered the 1973 flood the largest since 1927. The Corps' official estimate for the maximum flow is 2,005,347 cubic feet per second.[55] As discussed in the opening section on terminology, a flood's volume in cubic feet per second comes from the area of the river channel's cross section multiplied by the measurement of the flow velocity. Since the Mississippi is constantly reshaping its bed, especially during flooding, the cross section at any given point remains in flux. And like many floods, the 1973 flood peaked or crested multiple times at different points along the river's length. The NOAA Southern Region Headquarters reports a peak flow past Baton Rouge of 1,381,000 cubic feet per second on May 13, a figure which does not include the approximate 500,000 cubic feet per second passing through the Old River control structures.[56] Another figure came in 1980 from John R. Harris, with Louisiana State University's Department of Civil Engineering, who estimated a maximum flow in May (day unspecified) of 2,041,000 cubic feet per second at Old River and 1,498,000 below Old River, indicating a rate of 543,000 cubic feet per second captured by the low sill and overbank structures.[57] Louisiana State University faculty members Rafael G. Kazmann and David B. Johnson, whose 1980 study of the possible effects of a Mississippi River channel change at Old River will be discussed in Chapter 7, estimated a total peak discharge of 2.3 million cubic feet per second.[58] Based upon these different estimates, the 1973 flood probably had a peak discharge just above Old

River of 1.8 to 2.3 million cubic feet per second. As stated in Chapter 4, a 1930s Corps of Engineers estimate for the 1927 flood was 2.4 million cubic feet per second.

To cope with this much water, the floodways of the Corps' Mississippi River and Tributaries Project provided a managed alternative to the horrendous crevasses of 1927, drawing the main-channel flood down to a safe level by the time that it rolled past New Orleans. The unplanned activation of the Morganza Floodway, however, serves as a reminder that the Corps certainly didn't have the flooding Mississippi under control in 1973. But for the low-sill dilemma, Morganza probably wouldn't have come into operation until the high flow in May. Once activated, the Corps kept the Morganza Floodway open for fifty-eight days, finally shutting down the structure in the second week of June. The Bonnet Carré spillway remained open for a record seventy-five days and reportedly raised the surface of Lake Pontchartrain three feet above its normal surface level. According to the Corps' official count, twenty-eight people died in the 1973 flood, a heartbreaking statistic that is mitigated only in comparison to the hundreds of fatalities that occurred during the 1927 flood. Fortunately for the thousands of residents in the West Atchafalaya Floodway, the fuseplug levee at its northern end held and the Corps did not have to activate this outlet.[59]

The receding floodwaters finally gave the Corps a chance to assess the damage to the low sill. With apologies to Cecil B. DeMille and Charlton Heston, the engineers beheld the Mississippi's mighty hand. Figure 6.1 shows the extent of the damage to the structure's foundation due to scouring at the foot of the south wing wall. The eddy blasted out an oblong hole around 55 feet deep stretching more than 260 feet beneath seven of the low-sill's eleven gate bays. Nothing remained of the fallen wing wall, which the river pulverized and swept away through the structure's gates. Also shown in Figure 6.1 is the decade-old scour damage in the tail bay, refilled with riprap several times and washed out again in 1973 to depth of about 50 feet. The two holes were within 150 feet of each other. Had they joined, the low sill would have been undermined by the Mississippi's turbulence.[60] Would the structure have collapsed like the wing wall? Perry Gustin maintains that the pilings anchoring the low sill would have held it upright, even as the river roared beneath and invariably around the concrete ruin.[61]

CHAPTER 7

The Warning

The Old River region "is not the place where the works of man have a great prospect of a long life," wrote Louisiana State University professors Raphael G. Kazmann and David B. Johnson in their bluntly titled 1980 report, "If the Old River Control Structure Fails? The Physical and Economic Consequences."[1] The 1973 flood lifted the veil on an alarming situation that Kazmann and Johnson, and their LSU colleague, John R. Harris, took seriously enough to explore in detail for the Louisiana Water Resources Research Institute.[2] For residents on the Mississippi River below Old River and in the Atchafalaya Basin, the findings by Kazmann, Johnson, and Harris do not make comforting reading. At the outset of their report, they stated their "most important single conclusion":

> In the long run the Atchafalaya River will become the principal distributary of the Mississippi River and . . . the current mainstem will become an estuary of the Gulf of Mexico.[3]

This change in the course of the Mississippi River could be expected despite the Corps of Engineers' best efforts. What moves the channel diversion from a possibility to a certainty is the authors' qualifying expression "in the long run." The Mississippi River has time on its side. And as General Noble noted in reference to the 1973 loss of the low-sill's wing wall, "The punishing effect of water is always easy to underestimate."[4] No matter how sound the engineering, the

Mississippi will always discover and exploit weak points in any control structure's design.[5]

When the three LSU professors undertook their study, public memory of the 1973 flood was still vivid. Before the next major flood event, those in charge of municipalities and industries in the threatened areas—below Old River along the Mississippi and Atchafalaya rivers—needed basic facts and figures regarding potential losses and clear recommendations on how best to avoid or lessen the impacts. For Kazmann, Johnson, and Harris, the need to acknowledge the threat and make preparations was "only prudent."[6] They decided that the best way to proceed with educating the public was to first present a scenario for how the disaster might play out. The 1973 flood provided the authors with a believable opening scene: the failure of the low sill.

Their cautionary tale begins at the crest of a fictional flood along the lines of those experienced in 1927 and 1973. When the professors collaborated on their study, the auxiliary control structure and hydroelectric generator (discussed in Chapter 8) had not yet been built. For thirteen nervous years after the 1973 flood, the damaged low sill and its neighboring overbank structure were all that stood between the Mississippi River and a new course through the Atchafalaya Basin. It was during this interim that Kazmann, Johnson, and Harris produced their report, and they could easily envision a moment when the low-sill's luck might run out.[7]

In their doomsday scenario, the imagined flood overwhelmed the low sill, and the ruined structure partially blocked the Mississippi floodwaters rushing against it. This time, the use of the Morganza Floodway didn't mitigate the pressure on the Old River control area. As the river brought more and more force to bear on the critical diversion point, a crevasse decimated the mainline levee immediately south of the low sill. This rapidly widening fissure became the head of the Mississippi's new channel. Within a short amount of time, the avulsion reversed the 70 percent/30 percent division maintained by the Corps. From a flood stage discharge of around 800,000 cubic feet per second, the Atchafalaya River doubled in size to 1.6 million cubic feet per second, more water than the Atchafalaya Basin endured in 1927. Below the diversion point, the Mississippi's channel began filling with sediment.[8]

The authors wrote their Mississippi River channel-diversion narrative in the form of a fictional history, as if they were describing a recent chain of events stemming from the destruction of the low sill. For the rest of the scenario, I will use their predictions of what would likely occur in the Atchafalaya Basin and on the Lower Mississippi should today's Old River Control Complex fail, updating the picture where necessary to cover developments over the last thirty-five years.

◆ ◆ ◆

THE NEW MISSISSIPPI RIVER would impact hundreds of thousands of people in Atchafalaya Basin, and flood some three million acres, soaking everything between Bayou Teche and the Mississippi's old channel. Spreading out to claim the floodplain, the water would inundate all of Morgan City; Berwick; Melville; Krotz Springs; and parts of Franklin, Houma, and Thibodaux. Morgan City would be partially entombed in sediment. Ring levees at Melville and Krotz Springs might provide some protection, but the loss of electrical power, natural gas, and clean water would make these towns inhospitable islands at best. The region could expect widespread homelessness and unemployment. Until the Corps could build new levees and structures to gain some semblance of control over this version of the Mississippi—a process that might take decades, widespread flooding to varying degrees would become an annual nightmare.[9]

Beyond the loss of human life and the destruction of homes and property, Kazmann, Johnson, and Harris warned of the vulnerability of pipelines and transportation routes crossing the Atchafalaya Basin. Natural gas pipelines in profusion span the lowland where the Mississippi will come to establish its new channel, and natural gas currently generates 75 percent of Louisiana's electricity. Major interstate pipelines like Texas Gas Transmission move natural gas across the Atchafalaya to provide power for far-flung customers in midwestern and northeastern states. These conduits are vulnerable to scouring and destruction by the force of the new river and the tons of debris the current will be carrying southward. Although some of these natural gas pipelines currently hang suspended over the Atchafalaya River and others are buried beneath the riverbed, they would probably not escape the stream's transition to one of the world's largest waterways.[10]

Alon USA Energy operates an oil refinery at Krotz Springs. Two main petroleum pipelines, Colonial and Plantation, cross the Atchafalaya Basin carrying refined oil from sources along the Gulf Coast to numerous Eastern Seaboard consumers. The two pipelines run from Houston through Washington, DC, to the Philadelphia–New York City area. Colonial passes through Krotz Springs and Plantation through Morgan City. The temporary loss of one or both of these pipelines would cause major problems for numerous busy highways and airports.[11]

The four main highways across the Basin—Louisiana Highway 1 (Marksville to New Roads), US Highway 190 at Krotz Springs (Opelousas to Baton Rouge), Interstate 10 (Lafayette to Baton Rouge), and US Highway 90 at Morgan City (Lafayette to New Orleans)—could sustain enough damage to close them down. The 1973 flood scoured a hole more than 185 feet deep near the I-10 bridge over Whiskey Bay, 30 feet deeper than the bridge's center-line pier foundation. Luckily, the scour hole developed downstream from the bridge; however, the damage revealed a potential problem with bridge piers in a larger flood event.[12]

Also, four railroad lines span the danger zone. The Kansas City Southern (formerly the Texas and Pacific) from Alexandria through Baton Rouge to New Orleans parallels Highway 1 across the Atchafalaya and crosses the Morganza control structure. The Union Pacific track leaves Alexandria and crosses the Atchafalaya on the steel trestle at Melville, continuing through Grosse Tete to New Orleans. Another Union Pacific line from Opelousas parallels Highway 190 through Krotz Springs to New Orleans. Across the south rim of the Basin, the combined Burlington Northern Santa Fe/Union Pacific/Amtrak (Sunset Limited) leads from Lafayette through Morgan City to New Orleans. As with the highways, scouring could potentially undermine the rail trestles.[13]

◆ ◆ ◆

DOWN THE MISSISSIPPI'S OLD CHANNEL, the story is quite different, and equally grim. Out of twenty major electrical generators between Baton Rouge and New Orleans, seventeen are currently powered by natural gas. The reliability of electrical and gas utilities services would depend upon how many natural gas pipelines in the Atchafalaya Basin remain operable. Numerous riverside communities and industries are vulnerable, as is the New Orleans metropolitan area, where over four hundred thousand people could be exposed to potential outages.[14]

The availability of fuel for automobiles, airplanes, and other forms of transportation could also be a serious concern. At least ten major oil refineries hug the Mississippi River between Baton Rouge and New Orleans, along with three oil import sites. Under normal conditions, pipelines from these refineries carry oil products throughout the eastern United States. The electrical outages that will plague much of the region would hamper, to some extent, the flow of gasoline and diesel to local convenience stores and other fueling stations.[15]

In the Kazmann, Johnson, Harris scenario, 30 percent of the Mississippi's discharge continues to flow down its old channel. (The authors acknowledge that this is a conservative estimate; the Atchafalaya could capture the entire Mississippi.) With the river swollen by the flood, the change in the volume of water moving past the New Orleans levee would not become noticeable for several weeks, and the transition to the new, diminished Mississippi flow could take up to three or four months.[16] During this period, unless the New Orleans area has made the necessary advance preparations (discussed below), emergency evacuations should be well underway. In tandem with the exodus, a massive effort to truck fresh water to all the city's neighborhoods should be in progress to offset the inevitable loss of the Crescent City's primary source of drinking water. Currently, the New Orleans intake pumps draw about 210 million gallons per day from the Mississippi. With its reduced flow, the river will no longer be able to prevent Gulf saltwater from eventually threatening the city's water system.[17]

At the Mississippi's mouth, some one hundred river-miles below New Orleans, the "thalweg,"[18] the deepest part of the channel bed, fluctuates between fifty and one hundred feet below sea level. Here the Gulf water, being heavier and denser than the freshwater, spills into the river channel to form two layers of water flowing in different directions. Moving north beneath the freshwater that is flowing seaward, the nose of the saltwater intrusion takes the shape of a wedge with its leading edge snaking along the river bottom. Most of the time, the Mississippi's tremendous force prevents the saltwater wedge from intruding more than about twenty miles, to the vicinity of the town of Venice. Beyond Venice, the Mississippi's channel deepens considerably in places, reaching depths of nearly two hundred feet below sea level between Chalmette and New Orleans. In unusually dry years when the Mississippi's discharge diminishes, saltwater can follow the thalweg all the way to New Orleans and beyond. According to Harris, the maximum saltwater intrusion occurred in October 1939, when the wedge reached the town of Norco, around fifteen miles north of New Orleans. That year, the Mississippi's discharge fluctuated between seventy-five thousand and one hundred thousand cubic feet per second for thirty days in a row.[19]

The Mississippi's average annual discharge past Vicksburg is about 565,000 cubic feet per second.[20] Using the Kazmann, Johnson,

Harris scenario's distribution at Old River, approximately 395,000 cubic feet per second (70%) would pour into the Atchafalaya Basin leaving around 169,000 (30%) flowing past New Orleans. If the Mississippi would hold this discharge rate, the Crescent City and the communities and industries that depend upon the river could possibly adjust to a new status quo. But the river's annual rise and fall means that dangerous periods of low water would be coming.

The Mississippi's normal dry-season low flow rate is approximately 250,000 cubic feet per second at Vicksburg and Natchez. If 70 percent is diverted at Old River, a much diminished Mississippi River will carry only about 75,000 cubic feet per second to New Orleans during the low water season (August–January). A discharge this low over a period of weeks would bring saltwater into the New Orleans water system. If this low discharge lingers for six months, the Gulf water could reach Baton Rouge, 230 miles above the Mississippi's mouth. At New Orleans, the leaders of Jefferson and Orleans parishes would face a tough decision. They could protect public infrastructure and private property from saltwater corrosion by shutting down the intake pumps, leaving the city without running water. The other option, to continue pumping the salty water for sanitation and firefighting, would corrode the pipes and appliances throughout the city.[21]

Besides New Orleans, the Mississippi provides most of the fresh water for industries and communities south of Baton Rouge, including the Bayou Lafourche distributary from Donaldsonville to Thibodaux. When Kazmann, Johnson, and Harris conducted their study in the late 1970s, total Mississippi River water usage here was around six billion gallons per day. In 2010, Mississippi water consumption from Baton Rouge to New Orleans and Donaldsonville to Thibodaux was just over five billion gallons per day. Louisiana Department of Transportation and Development statistics indicate a decrease in industrial usage, while public use has risen slightly. Overall, the industrial trend has been toward less use of surface (Mississippi) water and increased reliance on aquifers.[22]

According to the 2010 water usage report, the big water users were power generators, which accounted for 62 percent of all the Mississippi water consumed below Baton Rouge. Industrial usage, at 32 percent, came in a distant second. These facilities include chemical plants, synthetic rubber and plastics manufacturers, petroleum refineries, food processors, and wood and paper product producers. Mississippi water for public supply in 2010 was just 4 percent of the total consumption, with negligible use by livestock and irrigation. After the Mississippi falls to its new, diminished-flow levels, industries using the river for "once-through cooling" before flushing the heated water out into current could begin to experience problems. With a slowed-down current, an industry's intake pumps would risk bringing some of the heated water back into its system. As the saltwater wedge pushes upriver, brackish water would pose corrosion problems for all users. Electrical generating stations and riverside industries would need up to two years to refit their systems for saltwater. Harris speculated that many of these businesses would elect to move elsewhere rather than make the adjustment to the new regime. Kazmann and Johnson noted that Lower Mississippi industries and communities might gain a brief window of time by the release of water from upstream reservoirs, but this action would trigger new problems with hydroelectric dam operations, water supply and recreation for communities around the reservoirs, the safety of communities below the dams, and wildlife management concerns.[23]

◆ ◆ ◆

KAZMANN, JOHNSON, AND HARRIS did their best to come up with cost estimates for some of the losses that would result from the broad categories of transportation disruption, flood damage, and pipeline failure. Their 1970s figures, even when updated to 2016 dollar values, appear quite modest and unrealistic, betraying the difficulty in quantifying a disaster of this magnitude. Indeed, the authors conceded that they were unable to assign monetary values to things like refugee relocation costs, long-term disruption of oil and gas production, problems caused by extended periods of electrical brownouts and blackouts, loss of both wild and domestic animals in the Atchafalaya Basin, dredging on the diminished Mississippi to allow barge navigation, disruption of agricultural activities, and the impact on the Gulf Coast seafood industry.[24] Likewise, the domino effect stemming from the loss of jobs in all south Louisiana industries, including shipping, education, manufacturing, timber, oil and gas, tourism, and health care would be impossible to calculate. If a silver lining can be found among the destruction, Kazmann and his colleagues predict that, while the oyster and shrimp beds may not survive the abrupt changes to water temperature and salinity, the seafood industry will recover over the long term. After all, these creatures have been adjusting to the Mississippi River's shifting outlets along Louisiana's coast for millennia.[25]

Unlike the coastal shellfish, the human population on the Lower Mississippi River and in the Atchafalaya Basin has the ability to foresee the possibility of disaster and make preparations. This is what Kazmann, Johnson, and Harris hoped to accomplish with their report. Thirty-five years later, only one of their recommendations stands completed. The report's top priority went to the construction of the Corps' auxiliary control structure at Old River (discussed in the next chapter), which was already in the advanced planning stage when "If the Old River Control Structure Fails" appeared. While the Old River Control Complex is now a more formidable deterrent to a Mississippi River channel jump, the LSU professors did not view the auxiliary structure as a permanent solution. If they had, they would not have taken the time and effort to research and present a number of other recommendations that could mean the difference between inconvenience and catastrophe. None of their preventive measures are easy or cheap, but all are within reach and make good sense in a region that is visited frequently by floods and storms.

For the Atchafalaya Basin, the disaster-preparedness suggestions offered by the report are nothing less than prudent advice for the

Mississippi Valley's largest floodway. After all, the Basin is where one-half of all Mississippi River floods, both modest and colossal, will go. In light of this, the foundations of all of the highway and railroad bridges across the Atchafalaya Basin need to be fortified, based upon the dangerous scouring that occurred near the I-10 support pier in 1973. The report recommends using steel pilings to anchor bridge pier foundations and injecting solidifying material such as grout to a depth of about 100 feet around the structures. Towns in the path of the flood should protect themselves with ring levees. Those communities that already have ring levees, like Melville and Krotz Springs, should improve them. Likewise, cities and towns that were flooded in 1927, even though they may be currently outside of the floodway's guide levees, should invest in ring levees. To help prevent the region-wide power outages that would follow pipeline destruction in the Basin, Kazmann and colleagues urged the strengthening of the pipeline crossings and cooperation between pipeline companies to allow alternate routing for all gas lines. As a further precaution, pipeline companies should band together to place one or more large pipelines deep beneath the Atchafalaya River, at depths of at least 250 feet, to be shared in an emergency to keep the gas moving. Finally, at the bottom of the floodway, Morgan City is vulnerable during any sizable flood. The report suggests the construction of additional Gulf outlets to augment the existing Atchafalaya River and Wax Lake outlets.[26]

On the Mississippi River side, a secure supply of freshwater will be the main issue. If the course of the Mississippi River changes, Kazmann, Johnson, and Harris predicted "direct political confrontation" between communities and industries that rely on the main stem of the Mississippi for water and surrounding communities that have plenty of water from other sources and wish to hang on to it. The solution, of course, is the development of alternative water resources before the time of crisis.[27]

In 1981, possibly inspired by the Kazmann, Johnson, Harris report, the US Geological Survey and the Louisiana Department of Transportation and Development began investigating the possibility of pumping water for public consumption from the system of aquifers beneath the New Orleans metropolitan area. Industries have been withdrawing water from these reservoirs since the 1940s, but the aquifers have not been tapped for drinking because of the water's unappealing, yellowish color. Despite the Crescent City's exposure to floods, hurricanes, and pollution from upriver oil spills and industrial mishaps, no one gave these local aquifers serious consideration until the horrific water shortage in New Orleans following Hurricane Katrina in 2005.[28]

If the New Orleans aquifers can be accessed in an emergency, they do not hold enough fresh water to be a long-term solution. Harris, in his part of the Kazmann report, "Alternate Water Sources for the Baton Rouge–New Orleans Industrial Corridor," noted that a good supply of fresh surface water and groundwater is available fairly close to New Orleans, north of Lakes Maurepas and Pontchartrain. Here, potential water sources include the rivers Amite, Tickfaw, Tangipahoa, Tchefuncte, Bogue Falaya, and Pearl. According to geographer Craig E. Colten, the New Orleans water works company investigated the possibility of drawing water from the rivers to the north in the late nineteenth century and found the cost prohibitive. Use of the surface water and groundwater north of Lake Pontchartrain is still feasible, but both sources present problems. The surface streams are subject to dry season low flows that limit their ability to provide a consistent supply. Likewise, aquifers can be depleted through continued use and lead to subsidence of the land above them. Because saltwater underlies the freshwater aquifers in this area, removal of the freshwater opens the cavities to saltwater intrusion. In recent years, increasing industrial use of groundwater in the Baton Rouge area may pose a threat to aquifers currently serving public needs. However, Harris suggests that, with adequate planning, aquifers and streams can be used together to provide a reliable, continuing source of fresh water. The key is to offset the amount of water drawn from a groundwater source by injecting an equal amount of fresh water to maintain the aquifer's equilibrium. In this scheme, the aquifer becomes an underground water storage facility. In Harris's view, the West Pearl River, around thirty-five miles east of New Orleans, is a potential source of fresh water for the aquifer storage strategy.[29]

The two branches of the Lower Pearl River empty into the Gulf about three miles east of Lake Pontchartrain. Like the Mississippi, Gulf saltwater moves up the two Pearl River channels during dry season low flows, so the head of a proposed pipeline to New Orleans would have to be located a sufficient distance upriver. Harris found that in the driest seasons, the West Pearl's saltwater wedge can extend about six miles to the town of Pearl River and suggests placing the pipeline intake just above this point. Because of the West Pearl's seasonal rise and fall, the amount of water available to New Orleans would vary during the year. To avoid introducing surface pollutants into the groundwater source, Pearl River water must be treated before being injected into the aquifer. Harris noted that purification of the Pearl water could be done at the New Orleans treatment plants or at treatment plants installed at the aquifers. Harris also pointed out that using aquifers for fresh water storage can circumvent some of the problems associated with surface reservoirs, such as evaporation, pollution, and siltation.[30]

Given the Crescent City water supply's long-standing exposure to pollution from upriver oil refineries, industries, agricultural chemicals, and barge traffic, serious consideration of alternate water sources makes sense regardless of the Mississippi's channel diversion threat.[31] Despite its potential to provide a sustainable alternative water source, the city of New Orleans never gave serious consideration to Harris's Pearl River–aquifer strategy.[32] Today, this kind of long-distance water transport within Louisiana is more restricted than in the late 1970s. In recent years, new state laws have placed strict regulations on the use of surface water and groundwater by individuals and entities other than the landowners who own property connected to these sources. Other options discussed in the Harris report include desalination, which was prohibitively expensive three decades ago and remains so today, although twenty-first-century drought conditions experienced in the western United States have kept this hope alive.[33] Harris also suggested recycling of sewage water for public use, which has helped alleviate water shortages in Europe and California.[34]

Kazmann, Johnson, and Harris hoped their study would awaken New Orleans, the Louisiana legislature, and Congress to the possibility of a map-changing diversion by the Mississippi River. The authors argued that the threat was "not a parochial problem . . . but will directly or indirectly affect a sizable proportion of the population of the United States."[35] But instead of spurring state and federal forces to action, few people, aside from engineers and academicians, read "If the Old River Control Structure Fails." The big, life-changing floods on the order of 1927 and 1973, like the biggest hurricanes, come too seldom to compete with more immediate priorities. For its part, Congress was backing the expansion of the Corps' control installation at Old River with a new structure to restore peace of mind and confidence in the status quo. Optimistic stakeholders in the region still hope that if enough concrete is poured between the Mississippi and the Atchafalaya, the scenario envisioned thirty-five years ago by the three LSU professors will remain locked in the future.

CHAPTER 8

Preparing for Uncertainty

Good luck plus General Noble's bold activation of the Morganza Floodway prevented the Kazmann, Johnson, and Harris scenario from becoming a reality. Coming more than forty years after the 1927 flood, the flood of 1973 showed a new generation of Americans what the Mississippi River is capable of doing. As the high water receded, Noble called for the construction of a new control structure at Old River, but he noted that the 1973 flood revealed changes in the Mississippi's channel from the time the low sill came into operation. This meant that hydraulic testing at the site and with models at the Corps' Waterways Experiment Station at Vicksburg would have to precede any new installation. In the meantime, the low sill needed extensive repairs.[1]

With no other options available, the low sill had to continue operating at full capacity while the engineers did their best to patch the wounded edifice. Corps engineers knew that scouring had occurred in both the fore bay and tail bay but were unsure about the extent of the damage underneath the structure. To find out, they drilled a hole down through the low sill and lowered a television camera into the depths. The camera showed that the floodwaters had washed away the earth beneath the structure, exposing the pilings. When General Noble looked at the television monitor, he saw fish swimming around instead of solid ground and concrete.[2]

While the new wing wall was taking shape, flotillas of barges loaded with riprap arrived to fill the fore- and tail bay holes. This approach might offset the damages on either side of the low sill, but there was no practical way to place the riprap boulders into the cavity the river had hollowed out underneath the structure. The best solution for repair beneath the low sill was to inject grout, a mortar compound treated to solidify in water, into the void around the pilings. According to Perry Gustin, the Corps experimented with grout mixed with shavings of fiberglass and metal and decided to use the mixture containing metal.[3]

While the underwater work progressed, crews assessed the wear and tear on the overbank structure and the low sill's upper works. During periods of seasonal low water, the Corps modified the low sill's gates to allow better control of flow through the structure. The addition of a second hundred-ton gantry crane added versatility that was lacking in the 1973 emergency and facilitated the work on the gates. The low sill's original design included piezometers placed at critical points to measure water pressure on the structure. The engineers replaced the old sensors with new instruments during the post-1973 overhaul.[4]

Despite a decade's worth of repairs, the low sill's foundation remained permanently damaged. The Corps cautiously raised the head differential that the structure could safely handle from eighteen feet right after the flood to a modest twenty-two feet, well below the low sill's original design head differential of thirty-seven feet. The engineers agreed that the reduced capability would suffice for operations at normal river levels but that the structure probably could not withstand the forces that a sizable flood might bring to bear on it. John D. Higby Jr, in a 1983 study for the Colorado Water Resources Research Institute, also warned of the danger from more barge mishaps. A loose barge stuck in one of the low sill's gates during a flood might force a gate closure and raise the head to a dangerous level. In fact, Corps studies showed that the Mississippi River channel's thalweg tracing the river's swiftest current had moved closer to low sill, increasing the danger of runaway barge collisions.[5] Preventing barge collisions, raising head differential, and improving gate operations were some of the low sill's design

problems that guided the Corps' planning for the next control
structure at Old River.

◆ ◆ ◆

BIGGER AND STRONGER than the low sill, the functionally titled
auxiliary structure gives the impression that it can control any of the
earth's mightiest rivers (see Figure 8.1). Taking the lesson learned in
1973 to heart, the Corps secured the entire structure with steel pilings, including the sweeping 80-foot-high and 178-foot-long wing
walls. Six 62-foot-wide gate bays across the structure's 442-foot
length match the low sill's depth with a bottom sill elevation of 5
feet below national geodetic vertical datum (which replaced mean
sea level in the 1970s).[6] For protection against the outflow scouring
that plagued the low sill, the stilling basin behind the gates is over
twice the length of low sill's stilling basin. To calm the water rushing
through the structure, the stilling basin is armored with two rows
of 15-foot-high, concrete baffles and a 12-foot-high end sill leading
into the outflow channel. On the Mississippi River side, the new
structure's approach channel is approximately 8,000 feet long, a little over three times longer than low sill's approach channel[7] (see
Figure 8.2).

Perhaps the most striking aspect of the auxiliary structure's
appearance is its seven forty-foot-high towers atop the span. Quite
different in design from the Corps' other Mississippi River control
structures, each tainter gate, as they are called, has its own lifting
mechanism comprised of tower-mounted cable and drum hoists,
which replace the cumbersome gantry cranes on the other structures (see Figure 8.1). The curved, seventy-five-foot-high steel gates
can be operated quickly compared to vertical lift gates, but their
design requires a long trunnion or projection on the downstream
side fastened to a shaft to allow the gates to rotate up and down.
Importantly, the auxiliary structure has a design head differential
of fifty-five feet compared to the low sill's original thirty-seven-foot
head capability.[8]

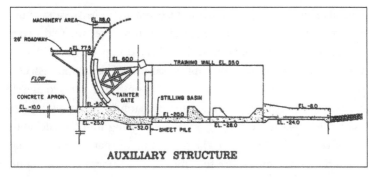

AUXILIARY STRUCTURE

Figure 8.1 The auxiliary control structure shown in cross section. Fairless, "The Old River Control Project," Figure 5. Image courtesy of the American Water Resources Association.

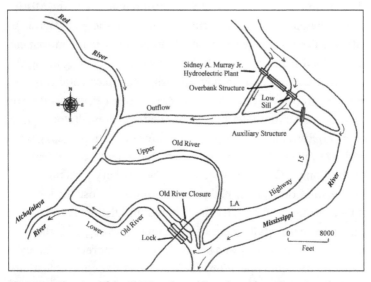

Figure 8.2 Plan view of the Old River Control Complex with auxiliary control structure and hydroelectric plant.

While the engineers planned the auxiliary structure, sediment studies at the Waterways Experiment Station convinced the Corps' New Orleans District to set a specific goal for the distribution of bed-load sediment between the Mississippi and Atchafalaya Rivers. Previously, the Corps had viewed the bed-load distribution in general terms, hoping to maintain the diversion of Mississippi mud that existed before the control structures took over the Old River connection. The hydraulic models indicated that the low sill was diverting 44 percent of Mississippi's bed load. The auxiliary structure would increase that amount, allowing both structures to divert approximately 65 percent of the main channel's heavy sediment. The outlook was encouraging: more mud going into the Atchafalaya River would slow down the deepening of that river's channel and at the same time reduce the amount of sediment remaining in the Mississippi to aggrade the channel below Old River. Based upon the modeling results, the New Orleans District assumed that the structures permitted such fine-tuned level of control over bed-load distribution and set the goal at 35 percent remaining in the Mississippi's channel versus 65 percent diverting into the Atchafalaya. As later studies reveal, the Corps' ability to manipulate the movement of bed-load sediment through Old River was a modeling mirage.[9]

Construction began in July 1981 with a supplemental congressional appropriation, and the auxiliary structure began operating in 1986. In the nervous interim while the damaged low sill stood guard over Old River, the Mississippi raised three major floods (1975, 1979, and 1983), all of which were serious enough to require the operation of the Bonnet Carré spillway. When the auxiliary structure came online, the Corps finally had the opportunity to close the low sill's intake channel and dewater the battered installation, revealing for the first time the full extent of the damages wrought by the 1973 flood. More repairs followed before the Old River crew put the aged structure back into operation. The Waterways Experiment Station's models indicated that the auxiliary structure would reduce the low sill's discharge by about 35 percent, which would in turn reduce the perennial scouring behind the smaller structure. The models also predicted that operation of the auxiliary structure would be a deterrent to the low sill's barge collisions.[10]

When the new structure went into operation, the Old River trouble spot once again stood ready for the project flood. Corps historian Martin Reuss remarked that the concrete and steel complex beside the Mississippi projects "durability and reliability; they are public works in a style that unmistakably marks the land."[11] With its majestic towers looming over the surrounding floodplain, the auxiliary structure certainly sets the "style," in Reuss's characterization. Ironically, the same flood that forced the Corps to build the new structure inspired another ambitious construction project at Old River, one aimed at harnessing the Mississippi's energy instead of fighting it.

◆ ◆ ◆

SIDNEY A. MURRAY JR. was one of the many curious visitors at the Old River Control Complex in the spring of 1973 to witness the battle between the Corps and the Mississippi River. Murray, the mayor of Vidalia, Louisiana, the riverside town some thirty-five miles north of Old River, came away from that experience with a visionary idea. By 1978, the mayor had convinced the Corps of Engineers of the feasibility of using the head differential at Old River, the sometimes precarious difference in elevation between the Mississippi and the Atchafalaya, to generate hydroelectric power. The water's fall from the Mississippi's level down to that of the Atchafalaya would supply the necessary force to spin turbine blades. As the project moved forward, it became a joint venture between Vidalia, the Corps of Engineers, and energy developer Catalyst Old River Hydroelectric Limited Partnership. Initially, the Corps considered the possibility of incorporating the hydroelectric facility with the auxiliary structure, which was then in the planning stage, but by the early 1980s, Vidalia and Catalyst were committed to a stand-alone generator on property immediately north of the overbank structure[12] (see Figure 8.2).

In 1985, with the auxiliary structure nearly complete, construction began on the generator site's inflow and outflow channels.[13] Termed "run-of-the-river," the hydroelectric generator would not impair navigation or the Corps' 70 percent to 30 percent water diversion ratio, but the army engineers were uncertain about how the new structure would affect the critical bed-load sediment distribution. While the project began to take shape, the Corps continued sediment testing with the models at the Waterways Experiment Station. Catalyst Old River Hydroelectric also experimented with scale models of the generator. In 1987, representatives from Vidalia, Catalyst, and the Corps traveled to Grenoble, France, to see a test model at the University of Grenoble.[14]

The world's largest prefabricated power plant emerged from Avondale Shipyards near New Orleans in the spring of 1989. For its voyage up the Mississippi to Old River, the colossus enjoyed a brief life as a ship, christened *The Merrimac* at her launch on April 8.[15] According to the Corps of Engineers, the $500 million "boat," measuring 456 feet in length and weighing 25,000 tons, remains "the largest vessel ever towed up the Mississippi past Baton Rouge."[16] Upon reaching its destination, *The Merrimac* became the Sidney A. Murray Jr. Hydroelectric Station, filled with 110,000 cubic yards of concrete that increased its weight by about 154,000 tons. Deep within the structure, the cascading Mississippi River drives eight twenty-four-megawatt hydrodynamic turbines. In the town's arrangement with Catalyst, Vidalia receives a percentage of the power generated by the hydroelectric plant, and utility companies purchase the remaining power.[17]

While contractors readied the plant to go into operation in late 1989, the Corps, Town of Vidalia, and Catalyst signed a memorandum of agreement that called for Catalyst to conduct a study of sediment transport in the Mississippi's Old River reach, including the movement of bed-load material through the hydroelectric generator. Ten years later, Catalyst produced a mountainous four-volume report, to which the Corps added a fifth volume with its evaluation of the Catalyst findings. Professionals assisting with the research included engineers from the Corps, Colorado State University's Engineering Research Center, the University of Iowa's Institute for

Hydraulic Research, and Mobile Boundary Hydraulics of Clinton, Mississippi. Simply titled "Lower Mississippi River Sediment Study," the team used both physical and computational models to determine the sediment transfer through the turbines. The project also studied historic discharge rates, river stages, and physical sediment sampling. When compiled, the mass of data indicated that the Mississippi's channel above Old River was aggrading or filling with sediment, possibility caused by displaced sand and silt still moving downstream from the Corps' 1930s channel cutoffs. The data also appeared to show that the worrisome reach below Old River had stabilized in 1990 and was no longer aggrading. This was good news to the Corps, as was the report's finding that the upper Atchafalaya also seemed to have stabilized and ceased degrading its channel. The main goal of the study, however, a clear assessment of the movement of bed-load sediment through Old River, eluded the experts' best efforts. In the next chapter, we will look at more recent efforts to grapple with this question. Part of the problem dates back to the consultant discussions in the 1950s with Einstein, Straub, and Bondurant and the difficulty in defining the Mississippi's "bed load."[18]

Today, the average distribution of water flowing through Old River is 75 percent hydroelectric, 20 percent auxiliary structure, and 5 percent low sill, allowing the generator to maximize its power output depending upon river conditions. A computer system controlled by the Corps monitors the flow amounts through the Old River structures to keep the Mississippi's diversion into the Atchafalaya at a steady 30 percent.[19] In 2010, approximately 85 billion gallons went through the hydroelectric generator's turbines. By comparison, the hydroelectric generator at the nearby Toledo Bend Reservoir on the Sabine River passed a little over 2 billion gallons during the same year.[20]

Despite the Sidney A. Murray Jr. Hydroelectric Station's quarter century of successful operation, the nation's energy-producing community still considers the Mississippi River an "untapped reservoir."[21] The twenty-nine locks and dams on the Upper Mississippi, above St. Louis, have attracted the attention of companies seeking to build conventional run-of-the-river hydroelectric plants. With no locks and dams on the Lower Mississippi, Old River represents the only opportunity for a conventional generator; however, a number of companies have begun to explore the use of stand-alone turbines in the Mississippi, either secured by pilings on the river bottom or housed on floating platforms. The muddy current averaging half-a-million cubic feet per second is a tempting source of power, but the Lower Mississippi may prove too rambunctious for these schemes to be practical.[22]

◆ ◆ ◆

LEVEES AND FLOODWAYS have handled the Mississippi's high water since 1927, but these devices cannot control the changes taking place on the bottom of the river. The evolution of the river's mainline levees is one way to track what is going on. In 1882, when John Ewens was at Old River, the standard levee size was 9 feet high and 53 feet wide at the base. By the early years of the twentieth century, levees' heights reached around 23 feet. After the 1927 flood, the Corps raised the levees to a prodigious 27 feet with a base width of 268 feet. By 1998, to stay ahead of higher and higher flows, the modern mound builders had lined the Lower Mississippi Valley with levees just over 30 feet high, resting on a pyramidal base 315 feet wide.[23] Rising flood stages also mirror the rising riverbed. At the Tarbert Landing gage, four miles upriver from the Morganza Floodway, the Corps measured the 1973 flood's volume at 1.3 million cubic feet per second and the gage registered a crest of 52 feet. In the 2011 flood, the same flow volume pushed the crest up to 55 feet on the Tarbert gage before the Corps activated the Morganza Floodway.[24] Likewise, the Red River Landing gage located on the Mississippi just below Old River has documented a climbing flood flow line: 1979, 59 feet; 1983, 60 feet; 1997, 62 feet; and 2011, 63 feet.[25] The Mississippi River floods are not getting bigger, but their elevation is rising.

As discussed in Chapter 1, the rising riverbed is part of the Mississippi's natural process as a sediment-hauling, alluvial river, but the Corps and other engineers also recognize the consequences of the artificial "improvements" made to the river over the last century. Repercussions from reservoirs, revetments, cutoffs, levee construction, and control structures may yet take decades to uncoil. The downstream movement of sediment caused by the 1930s meander bend cutoffs, which contributed to the rising riverbed below Old River, continued until the 1970s, and that process may still be active. In the next chapter, we will see that the installation of the low sill introduced significant changes in the movement of sediment through and past Old River that further contributed to the deterioration of the reach between Old River and Morganza.[26]

In response to the changing riverbed, the Corps raised the flow line for the project design flood, the imaginary deluge used in Mississippi River flood-control planning, but the volume of the project flood has remained much the same since 1927. Table 8.1 reveals a few minor changes to the predictive model. The amount of water predicted to stay in the Mississippi's main channel below Old River remains at 1.5 million cubic feet per second. Of that amount, the Corps is confident that the Bonnet Carré spillway can remove 250,000 cubic feet per second to keep the floodwater below the top of the New Orleans levees. Reflecting the increasing capacity of the Atchafalaya River, the volume diverting through the Old River control structures has increased from 500,000 to 680,000 cubic feet per second. In contrast, the amount to be shunted through the Morganza Floodway decreased by 50,000 cubic feet per second, perhaps a concession to the residents of Morgan City and the other communities at the other end of the Morganza pipe. The West Atchafalaya Floodway's share of the project flood saw a more dramatic decrease, from 350,000 to 250,000 cubic feet per second. Although the Corps originally designed the West Floodway as an integral component of the Atchafalaya Floodway, the 170,000-acre zone differs from the neighboring Morganza Floodway in one important respect: people live there.[27]

TABLE 8.1 Changes in Corps of Engineers project design flood volumes and distribution between the Mississippi's main channel and the three intake points of the Atchafalaya Floodway.

	Before 1945*	1951*	2012**
Mississippi R	1,500,000	same	same
Atchafalaya R	500,000	600,000	680,000
Morganza	650,000	600,000	600,000
West FW	350,000	300,000	250,000
	3 million	3 million	3.03 million

Sources:
* Latimer and Schweizer, *The Atchafalaya River Study*, Vol. 1, Appendix D, 28
** Camillo, *Divine Providence*, 10

When the Corps finalized the design of the Atchafalaya Flood-way, the Morganza corridor was largely farm and timberland with few residents. The comprehensive flowage easements the Corps purchased in the Morganza Floodway prohibit landowners from residing in the floodway but allow the continued use of the property for farming, timber removal, and other nonresident uses that do not conflict with flood control. In the West Atchafalaya Flood-way, part of the area that Major General Edgar Jadwin, the flood-way's creator, blithely referred to as swampland, family farms dot the countryside, and a string of towns abide alongside the west bank of the Atchafalaya River. From Highway 105 running beside the levee on the west side of the Atchafalaya River, travelers can see well-tended fields of sugarcane, cotton, soy beans, and milo. Cattle graze on the levee's slope across the road from trim, frame houses with close-mowed lawns, porch swings, ornamental fruit trees, shady pecan and oak trees hung with Spanish moss, silver propane tanks, and fluttering American flags. Because of the resident population, the West Floodway flowage rights are categorized as perpetual easements that allow the Corps of Engineers to use the land for flood control while the owners are allowed to live on the property and pursue the same activities permitted in the Morganza Floodway.[28]

Based upon 2011 figures, approximately 2,800 people live in the ring-leveed towns of Simmesport, Melville, and Krotz Springs beside the Atchafalaya River. Another 2,500 people live out in the floodway without flood protection. In all, the Corps estimates that the West Floodway contains around two thousand homes. Below the southern end of the West Floodway, where the Atchafalaya River levees end, the small town of Butte La Rose stands exposed without a rig levee. Along the northern border of the West Floodway paralleling Highway 1 just below Bayou des Glaises, General Jadwin's fuseplug levee provides a modicum of protection when the big floods inundate the Red River Backwater. So far, the Corps has never activated the West Floodway. On paper, the Morganza Floodway and the Atchafalaya River would both have to reach or even exceed their design capacities, a total of around 1.2 million cubic feet per second, before the Corps would consent to flood this picturesque piece of Middle America.[29]

◆ ◆ ◆

THE RISING FLOOD FLOW LINES also affect the probability that riverside communities once considered safe from high water might lose lives and property in a big flood. To hold down flood losses by keeping people out of flood zones, local governments and the Federal Emergency Management Agency (FEMA) determine the boundary between danger and safety with a biblical-sounding yardstick called the "100-year flood." Like the Corps' project design flood, the 100-year flood is a useful tool that is easier to define than understand. Where the project flood is a hypothetical volume of water measured in cubic feet per second, the 100-year flood is a probability, a flood that has a 1 in 100 chance of occurring in any given year. The statistical value of the 100-year flood makes our national flood insurance program viable, but the term can also give the false impression that such a flood will only happen once every 100 years.[30]

In his 1945 book, *Human Adjustment to Floods*, geographer Gilbert F. White framed the issue in common sense terms: "Floods are 'acts of God,' but flood losses are largely acts of man."[31] Beginning in the 1920s, geographers like White and Harlan H. Barrows, both associated with the University of Chicago, advocated what may seem obvious: people can avoid death and destruction from floods by staying out of flood-prone areas.[32] As discussed in earlier chapters, American Indians in the Mississippi Valley lived in harmony with the river, and their losses to floods were probably minimal. The problem became acute in the mid-twentieth century with expanding riverside communities. The Corps' floodplain maps held the answers to questions about historic flood elevations, prompting the Flood Control Act of 1960, which authorized the Corps to help state and local governments with floodplain planning.[33]

Even with access to the Corps' knowledge of flood histories and high flows, people continued to move into flood-prone places, putting private insurance companies in a quandary: compared to the total population, relatively few homes and businesses locate in floodplains, making premiums too expensive for many who really need flood insurance. In 1966, President Lyndon Johnson issued Executive Order No. 11296 denying federal assistance for construction projects without the completion of a flood hazard survey of the proposed site. The Corps worked with numerous other federal agencies developing guidelines to determine what places qualified as floodplain. As part of this process, the concept of the 100-year flood made calculating a place's flood hazard relatively easy using a flood frequency curve, a statistical graph plotting how often floods of certain sizes occurred at particular locations. Variations on the probabilities spawned other statistical intervals like 25-year and 1,000-year floods. One of the alarming conclusions in the 1980 report by Kazmann, Johnson, and Harris is their classification of the devastating 1973 flood as a mere 25-year flood.[34]

FEMA flood hazard areas drawn using frequency curves gave the federal government an organized way to set flood insurance rates. The National Flood Insurance Act of 1968 and the Flood Disaster Protection Act of 1973 (as amended) established FEMA's authority to administer a national flood insurance program in which the federal government partners with private insurance companies. The laws authorize the identification of flood-prone zones and encourage state and local governments to avoid development in these areas. Participation in federal flood insurance is not mandatory, but states or areas that opt out of the coverage risk denial of federal assistance for properties in flood hazard zones. Likewise, federally regulated lending institutions can only lend to insured properties in designated flood zones. The laws also authorize regular updates to account for changing flood conditions. Although the federal government has embraced flood frequency curves as a way to predict the future, H. C. Higgs, with the US Geological Survey, noted that statistical probabilities based on a small number of incidents could be misleading. Our database of major floods is quite limited, with reliable stage and volume figures reaching back only a little more than a century. Beyond that, recent archaeological studies indicate changing frequencies of major Mississippi River floods over time.[35]

Our historic window on the Mississippi Valley is indeed brief. A century is a long time to a human but only a tick on the Mississippi River's geological clock. Abstract concepts like the project flood and 100-year flood may give the false impression that the Mississippi and the valley through which it flows are dependably stable. If the river's bed is rising, the Corps' dredge boats are a plausible solution, as is the availability of dirt to raise levees. But more is happening with the Mississippi than a rising riverbed. The Lower Mississippi Valley is constantly responding to the earth's tectonic dynamics, land subsidence, and evolving climate.

◆ ◆ ◆

TECTONIC MOVEMENT and its effects usually go unnoticed without a major event like an earthquake. The Mississippi Valley's best-known example is the seismic activity around the southeastern Missouri town of New Madrid in 1811–1812. Three major earthquakes shook that area between December 1811 and February 1812 along a fault zone in the interior of the North American plate. The faults in the New Madrid area are far from the much more volatile region associated with the North American plate's collision with the Pacific plate along the California coast, yet they attest to pressure and stresses within the plate affecting the Mississippi Valley.[36]

Earthquakes caused by faulting or fracturing of the earth's crust are not uncommon in the Mississippi Valley, although most have occurred in the middle valley between Memphis and St. Louis. In 1930, people along the Mississippi River south of Baton Rouge at White Castle and Donaldsonville experienced an earthquake that made the ground tremble thirty-five miles away in Morgan City. Tremors also accompanied a mile-long crack in the earth in 1943 near Vacherie, Louisiana, upriver from New Orleans. These seismic events may have been caused by movement in the Baton Rouge Fault Zone, a series of east-west running faults extending eastward from the Baton Rouge area to the upper rim of Lake Pontchartrain and reaching depths of twenty thousand feet in places. The movement associated with the Baton Rouge Fault Zone is not dramatic. Its effects are usually limited to some displacement of roads and bridges and minor structural damage. South of Old River and in the lower Atchafalaya Basin, where deep alluvial sediments hide bedrock faulting, deep borings associated with petroleum and groundwater exploration can detect evidence of tectonic movements.[37]

Faulting associated with salt domes also impacts the Mississippi Delta region. While the Old River area is too far north to be directly affected by the salt formations on the Gulf rim, the Mississippi River between Vicksburg and Baton Rouge flows through a region

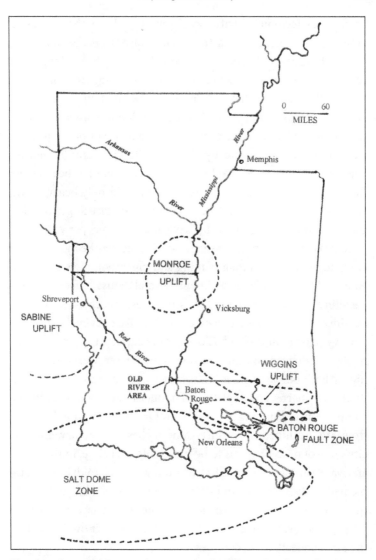

Figure 8.3 Areas of tectonic activity mentioned in the text. Based upon Saucier, *Geomorphology and Quaternary Geologic History of the Lower Mississippi Valley,* Figure 6.

of complex tectonic activity (see Figure 8.3). Here, three areas of active uplift or rises in the earth's surface caused by deeper forces are encroaching on the Mississippi's channel and may have a long-term influence on channel direction. Closest is the Wiggins uplift to the east of Old River, which may be nudging the Mississippi toward the Atchafalaya Basin. To the north, the broad Monroe uplift straddles the Mississippi Valley. The even larger Sabine uplift on Louisiana's western border may indirectly affect the Old River area by causing changes to the Red River. While the Monroe uplift has been active for around 136 million years, the Wiggins uplift only began rising about 50 million years ago. According to geologist Roger Saucier, the Wiggins uplift is growing by "several millimeters per year."[38]

In coastal Louisiana, active land subsidence may be more threatening to flood control than faulting and mushrooming salt domes. Ancient Mississippi River sediments several thousand feet thick are weighing on the underlying layers of limestone, chalk, and shale, bending the strata in a dip toward the Gulf like a diving board with a heavy person on the end. Closer to the surface, the ground normally waterlogged and thickened by annual river overflows is drying out and compacting due to the levees, human changes to age-old drainage patterns, and depletion of aquifers. Between the sediment compaction and sinking bedrock, New Orleans's elevation is falling from around nineteen to thirty-nine inches per century, and the city's rate of descent seems to be increasing over time. In the Atchafalaya Basin, geologists Ellis Krinitzsky and Harold Fisk documented sunken Indian mounds indicating active settling in historic times. Some of our modern mounds, the Atchafalaya Floodway's lower guide levees, are also disappearing under their own weight as the earth beneath them compresses.[39]

Along with sinking land, a rising sea level doubles the threat to south Louisiana. According to Tulane University geographer Richard Campanella, Louisiana has lost more than two thousand square miles of coastal marshes since the 1930s to this paired menace.[40] In 2004, a study for the National Oceanic and Atmospheric

Administration conservatively estimated that the Gulf is rising about 1.25 millimeters per year at Grand Isle, around fifty miles south of New Orleans. While this rate of rise might seem miniscule, it translates to nearly five inches per century, and research indicates that global sea level will accelerate during the twenty-first century.[41] For the Old River area, global climate change also brings uncertainty about future floods. The Corps' current operating policies and procedures recognize that extreme conditions may exceed past limits of climate variability in the decades to come.[42] In project planning, the Corps relies on the National Oceanic and Atmospheric Administration's National Environmental Satellite, Data and Information Service (NESDIS) for long-range climate change projections.[43]

To peer into the future of North American weather, NESDIS divides the country into eight regions and employs models that range from projections assuming no global effort to curb greenhouse emissions to models based upon steadily reduced emissions. For the Mississippi River, perhaps the most important region is the Midwest, comprising Minnesota, Wisconsin, Michigan, Iowa, Illinois, Indiana, Ohio, and Missouri. Temperature models for 2021–2099 show an increase of from 4.5° (F) to 8.5° (F) based on various emissions scenarios. Precipitation for the same period shows increases of from 1.5 percent to 4 percent for same period with a noticeable increase in the number of days with precipitation over one inch in the area at the upper end of Mississippi Valley made up of southeastern Missouri, southern Illinois, southern Indiana, and southern Ohio. The percent change for precipitation in the NESDIS scenarios is a comparison to amounts measured during a reference period from 1971 to 1999.[44]

The other important NESDIS region for the future of Mississippi River floods is the Great Plains states of Montana, North Dakota, South Dakota, Wyoming, Nebraska, Kansas, Oklahoma, and Texas. The projected range of temperature increase for this region from 2021 to 2099 is 2.5° (F) to 8° (F). The Great Plains models indicate a slight drop in annual precipitation during the century of from 1 percent to 2 percent.[45] The Southeast region, which includes the Lower Mississippi Valley states of Arkansas, Tennessee, Mississippi, and Louisiana, shows an increase in temperature very similar to that of the Midwest for 2021–2099; however, the models do not indicate a clear trend in precipitation. The projection for the Mississippi Valley is skewed somewhat by the inclusion of the southern Atlantic states. Models assuming no reductions in greenhouse gases indicate a much dryer scenario, while models taking global reductions into account seem to predict slightly wetter weather.[46]

Whether or not the NESDIS modeling projections prove accurate, these scenarios, along with the region's tectonic activity, describe a Lower Mississippi Valley that is anything but stable. To maintain control over the Mississippi's Old River diversion during the coming decades, the Corps' management of the auxiliary, low-sill, and hydropower structures will have to evolve with the changing consequences of human and natural forces. Despite the installation of the mighty auxiliary control structure, the Corps' mission at Old River still faces a formidable challenge. The problem is an old one. The movement of the Mississippi River's sediment is proving as difficult to understand in the twenty-first century as it was in the 1950s, when it defied Einstein's formulas.

CHAPTER 9

Awaiting the Project Flood

Most of the engineers who reviewed Catalyst's 1999 "Lower Mississippi River Sediment Study" were skeptical about the Old River diversion capturing 65 percent of the Mississippi's bed load, the target set in the 1970s by the Corps' New Orleans District. The nagging uncertainty about bed-load sediment that vexed Einstein, Straub, Bondurant, and the other engineers planning the Old River control structures remained shrouded in the Mississippi's mystique. The low sill was supposed to pull as much bed load as possible out of the river to prevent it from filling the Mississippi's channel downstream, but steadily rising stages with each flood showed that the system was not working. The Corps suspected that the problem involved both the changes over time in the river channel adjacent to Old River and the evolving hydrology, as new structures became part of the dynamic. In defense of the 1950s engineers, the Old River diversion in the twenty-first century is quite different from the place for which the low sill and outflow channel were originally designed.[1]

A study led by Ivan H. Nguyen with five other engineers at the Corps' St. Louis District Applied River Engineering Center, conducted between February 2010 and March 2011, indicates that the low sill probably captured around 30 percent of the Mississippi's bed load during its early years of operation, between 1964 and 1975. Compared to the Corps' expectations, the percentage is modest at best, but even that much bed-load diversion could not be sustained. The study by the St. Louis team indicated that the bed load pulled in by the low sill and later with the auxiliary structure decreased "dramatically" over time. Still more alarming is the finding that all of the bed-load material entering the structures' inflow and outflow channels stayed there. Although lighter-weight, suspended sediment seemed to move freely through the channels, none of the heavy sediment got to the Atchafalaya River. Since the installation of the hydroelectric plant, the problem has worsened. The hydroelectric intake channel, which now carried the lion's share of the Old River diversion, wasn't capturing any bed load according to the models. With the reduced flows in the channels of the low sill and auxiliary structures, bed-load sediment movement in the Old River Control Complex was at a standstill.[2]

For the Old River sediment study, Nguyen and his colleagues used an HSR model (Hydraulic Sediment Response), which is about the size of a ping-pong table as opposed to more conventional, large-scale hydraulic models. To replicate the desired river reach, the HSR laser scanner sculpts a polyurethane model from high-resolution aerial images of the river. Computer software regulates water flow through the model, and lasers track granular plastic sediment, as opposed to coal dust sediment normally used in large models. Because the quantity of the true bed-load component of the Mississippi's sediment at Old River remains unknown, the St. Louis team used an assumed amount based on the Corps' sediment samples from that part of the river. Along with the modeling experiments, the study included a review of the Corps' hydrographic surveys conducted since 1962 of the inflow and outflow channels for the low-sill and later the auxiliary structure. The surveys, measurements taken of the depth and shape of the channels, included hydrographic surveys for the Mississippi River channel through the Old River reach.[3]

The hydrographic surveys revealed a dramatic development in the outflow channel. Instead of performing like a flume conveying bed-load material from the Mississippi to the Atchafalaya, the diversion through Old River had evolved into an alluvial stream meandering within the confines of the 1,300-foot-wide channel, forming sandy point bars out of the captured sediment. Behaving like a "real" river, the point bars deflected the current against the opposite bank, causing erosion and widening the channel in places to around 1,600 feet. This meandering regime was probably in place by 1975.[4]

The Corps' hydrographic surveys also showed that the Mississippi's channel adjacent to Old River has changed considerably since the first surveys made in 1948. The thalweg tracking the path of the fastest current and deepest part of the channel shifted its course within the channel a number of times, repositioning the point bars where the bed load moved in the slower current. The low sill enjoyed a favorable bed-load situation in 1964 with the thalweg running along the opposite bank. Later surveys tracked the thalweg's migration across the channel to run alongside the intakes for the low sill and later auxiliary structure, moving the bed load out of reach near the opposite bank. The model studies indicated that when the hydroelectric plant began operating, its intake was too high and poorly positioned to take in bed-load sediment. Nguyen and colleagues also found through model experiments that structures such as weirs and dikes placed in the Mississippi near Old River would be unable to influence the current enough to shift the thalweg to a favorable path. The models permitted the team to try different combinations of open and closed structures, such as shutting down the hydroelectric generator while opening the low sill and auxiliary structures. None of the combinations showed any success in capturing bed load and clearing out the outflow channel.[5]

Back in September 1952, consultant Lorenz Straub had voiced his concern that the Old River outflow channel might fill with sediment and be unable to clear itself.[6] According to the 2011 report from the St. Louis District's Applied River Engineering Center, Straub's apprehension was coming true. Nguyen and colleagues concluded that a "fair amount" of bed load probably moved through the Old River diversion before the installation of the control structures. But after the construction of the low-sill and overbank structures,

Old River became "a sediment trap."[7] Based upon the Applied River Engineering Center's report, dredging the inflow and outflow channels to their original specifications would encourage a resumption of bed-load movement into the system, but the models predicted that the channels would refill over time, since little or no bed load would escape into the Atchafalaya.[8]

The models also implied that the problem was more than simply a matter of long-term, dredging maintenance. Nguyen and colleagues reported that the low-sill and auxiliary structures "could become totally buried with sediment during an extreme event."[9] The Mississippi's diversion into the Atchafalaya River could become blocked if the same extreme flood also rendered the hydroelectric generator inoperable, with severe consequences for both New Orleans and the Atchafalaya Basin. By coincidence, the ink was barely dry on the St. Louis District's Old River sedimentation study when the Mississippi River produced an "extreme event."

◆ ◆ ◆

THE CORPS' ESTIMATE for the maximum volume of the 2011 flood is 2.1 million cubic feet per second between Old River and Natchez. By comparison, the estimates for the volume of the 1973 flood in this same part of the river range from 1.8 million to 2.3 million cubic feet per second.[10] At Old River in 2011, approximately 672,000 cubic feet per second diverted through the hydroelectric plant and control structures into the Atchafalaya River, while a little over 1.4 million cubic feet per second rolled past Old River toward Morganza.[11] The Corps activated the Morganza Floodway for the first time since the 1973 flood and only the second time in its history and kept the spillway gates open for fifty-five days. The Bonnet Carré spillway operated for forty-three days. According to National Oceanic and Atmospheric Administration (NOAA) records, the gage at Red River Landing on the Mississippi just below Old River registered "an all-time record crest" of 63.39 feet on May 18.[12]

As Nguyen and colleagues predicted, the 2011 flood carried a lot of sediment into the Old River Control Complex. The structures weren't buried, but the mud infusion added tons of silt to the already choked artificial channels. With no time to mull over the St. Louis District's study, the Corps scheduled the first dredging operation at Old River since the initial construction of the low-sill and auxiliary structures. Between July 2012 and March 2014, cutterhead dredges reexcavated the two structures' inflow and outflow channels, working mostly during the low water season of midsummer through December.[13] Contacted in July 2015, Ivan Nguyen was unsure what the Corps' New Orleans District planned to do with the findings of the St. Louis District Applied River Engineering Center's report.[14] The extensive dredging operation following the 2011 flood may buy enough time for additional studies and strategic planning to address Old River's bed-load problem.

If the Corps was hoping for a respite from high water to begin dredging at Old River in the summer of 2012, the Mississippi more than complied. The drought of 2012 was the worst experienced in the Great Plains since 1895, even exceeding the heartbreaking Dust Bowl summers of 1934 and 1936. Although the Lower Mississippi Valley escaped the severe conditions, the Mississippi River's western and midwestern drainage basin bore the brunt of the dry weather. The great river's flow fell from a torrent to a trickle.[15] National Weather Service hydrologists reported river levels along parts of the Mississippi at thirty to fifty feet lower than the previous year's flood crests.[16]

Not surprisingly, the reduced flow brought the dreaded saltwater wedge snaking up the Mississippi's thalweg to threaten the New Orleans water supply. In response, the Corps refurbished a $5.8 million river-bottom dam used in 1988 and 1999 to block the wedge while allowing the southbound river water to flow over the top of the structure (see discussion in Chapter 7). The 1,700-foot-long barrier spans the Mississippi's channel near the town of Alliance, some sixty-four miles above the mouth and less than thirty miles below

New Orleans. Completed in mid-September, the dam successfully held the saltwater at bay until upriver rains eventually gave the Mississippi enough power to push the wedge back toward the Gulf.[17]

The Corps successfully handled the encroaching saltwater, but maintaining navigation between New Orleans and St. Louis proved to be a much more difficult challenge. The heavy sediment load left in the Mississippi's channel after the 2011 flood emerged as treacherous sandbars when the drought's effects persisted into the fall and winter of 2012. While the river fell, the Corps found itself in a quandary. Mississippi River shippers needed water releases from upstream sources to alleviate the low river conditions, but the reservoirs were also low, and the Corps had to maintain those levels for the local communities that depended upon them. Eventually, tow operators' hopes for relief focused on a faraway dam in South Dakota.

The Gavins Point Dam, at the southern end of Lewis and Clark Lake, is the southernmost dam at the bottom of a long string of dams climbing the Upper Missouri River past the vast Cheyenne River and Standing Rock Indian reservations into North Dakota and across Montana to the eastern slope of the Rocky Mountains. Every year in late fall, the Corps reduces the flow from the Gavins Point Dam from 37,500 to 12,000 cubic feet per second. Despite the navigation hardships on the Mississippi River below St. Louis and the outcry to keep the dam open, the Corps' operating procedures prevented the unscheduled releases. While the Corps remains an easy target for blame when natural disasters strike, the 2012 drought serves as a reminder that the army engineers simply cannot manage the nation's river systems to everyone's satisfaction at all times. In the Gavins Point case, water releases are coordinated with releases from the chain of dams further upstream along the Upper Missouri. The Corps must also consider Missouri River navigation and water use requirements downstream from the dam, and at the same time hold a sufficient water supply level in Lewis and Clark Lake.[18]

◆ ◆ ◆

WAS THE MISSISSIPPI'S EXTREME RISE IN 2011 AND FALL
IN 2012 a preview of coming erratic weather conditions sparked
by global climate change? Climatological observers are uncertain.
The beginnings of the 2011 flood are reminiscent of weather events
that preceded the big floods of the twentieth century, with heavy
winter snowfall in the Upper Mississippi Valley and along the Ohio
River, followed by ample spring rains in the same region.[19] In contrast, the 2012 drought seemed to come out of nowhere following
normal winter and spring precipitation. Termed a "flash drought"
by the NOAA because of its unpredictability, the record-setting dry
summer appears to be part of a recent drying trend in Great Plains
weather; however, the NOAA Drought Task Force stopped short of
linking the event with climate change.[20]

As discussed in this and previous chapters, the changing physical
characteristics of the Mississippi River, particularly its rising bed,
probably pose a more immediate threat to the Old River Control
Complex in the coming decades. The problems caused by the chan-
nel-filling south of Old River have been covered in some detail, but
we have not yet considered the potential effects of the rising riverbed north of the control structures. The 1999 Catalyst sediment
study found significant aggrading of the Mississippi River's channel
between Vicksburg and the vicinity of Black Hawk Landing, about
six miles upriver from Old River. Engineers working on the Catalyst report attributed some of this sediment accumulation to the
Corps' 1930s meander bend cutoffs and dislodged riverbed material moving downstream.[21] With river stages creeping upward from
one flood to the next, other places besides Old River could become
potential diversion points. The Mississippi River Commission first
took note of one such location of interest in 1882, at a rollercoaster
loop in the Mississippi's channel about five miles north of Black
Hawk Landing known as "Widow Graham Bend" (see Figure 9.1).

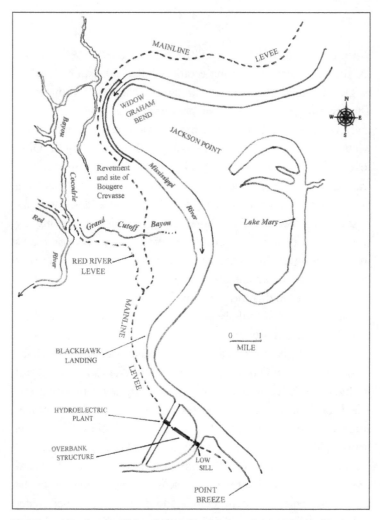

Figure 9.1 Map showing Widow Graham Bend, Jackson Point, and other locations mentioned in the text.

During the 1882 flood, John Ewens and his crew studying Old River (see Chapter 3) recognized the vulnerability of the west bank at Widow Graham Bend, where only five miles of floodplain separate the Mississippi and Red Rivers.[22] Beginning around 1859, the Mississippi occasionally overtopped the western end of the Widow

Graham meander loop near the community of Bougere, Louisiana. The overflows at this spot, which came to be known as the Bougere Crevasse, were probably a reaction to upriver levee construction, riverbed aggradation, and natural changes to the Mississippi's channel discussed below. Until the early twentieth century, the Corps' west bank levee system extending south from the mouth of the Arkansas River stopped at Bougere. With the upriver levees blocking numerous natural overflow spots, more water remained in the channel to escape at places like Widow Graham Bend. By the turn of the twentieth century, the Bougere gap had widened to about five miles before the Corps completed the levee line from Bougere down to Point Breeze near Old River in 1910.[23]

Closure of the Bougere Crevasse brought relief to west bank farmers who had been troubled by more than fifty years of overflows but angered landowners on the east side of the Mississippi. As long as the Bougere gap captured the Mississippi's high flows, the land across the river where E. H. and Mattie W. Jackson farmed cotton and corn remained undisturbed. After the Corps finished the levee at Bougere, the Mississippi's next overflow had nowhere to go except across some 1,100 acres of the Jacksons' farmland.[24] The Jacksons' property, known as Jackson Point, is a large point bar that formed as part of the Mississippi River's westward migration. With the river's help, the point bar had been growing steadily for well over a century following the natural cutoff in 1776 that created Lake Mary (see Figure 9.1).[25] As with the artificial cutoffs we've examined, the dramatic change in hydrodynamics can trigger profound repercussions. In this case, the Mississippi's energy turned westward to form Widow Graham Bend.

Between the 1820s and early 1950s, Jackson Point grew over two miles in length, filling in behind the migrating meander bend. When the Corps finally built the four-mile-long revetment to halt the Mississippi's westward progress, Widow Graham Bend had chewed the west bank to within about three hundred yards of the mainline levee (see Figure 9.1).[26] The shifting dynamics that contorted the Mississippi into Widow Graham Bend had another potentially

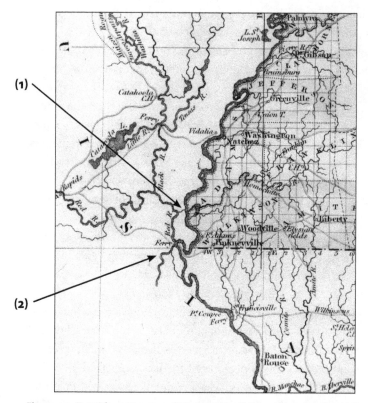

Figure 9.2 Detail from 1823 map of Mississippi by Fielding Lucas showing (1) Grand Cutoff Bayou connecting the Mississippi and Red Rivers north of the ferry site at (2) the Old River junction of the Mississippi, Red, and Atchafalaya Rivers. Lucas. (Map 1823) "Mississippi" [Baltimore]. Mississippi Department of Archives and History, Archives Library Collection MA/92.0062a. Courtesy of the Archives and Records Services Division, Mississippi Department of Archives and History, Jackson.

critical effect on the great river's course. Several nineteenth-century maps published between 1820 and 1878 show a nascent distributary known as Grand Cutoff Bayou flowing west out of the Mississippi about three miles south of the bend.[27] The bayou/distributary cut across the intervening floodplain to connect with the Red River (see Figure 9.2). Like Widow Graham Bend, the distributary formed following the aforementioned 1776 cutoff; Grand Cutoff Bayou does not appear on pre-1780 colonial maps of the area.[28] By the early twentieth century, construction of the mainline levee blocked the distributary, transforming it into a local drainage slough that still marks the path of this brief connection between the Mississippi and Red Rivers (see Figure 9.1).[29]

In 1913, during the deliberation over closing Old River, the Mississippi River Commission considered the close proximity between the Mississippi and Red Rivers at Grand Cutoff Bayou to be a favorable location for a lock.[30] This same stretch along the west bank could, under the right conditions, also become the Mississippi River's point of diversion into the Atchafalaya Basin. Standing in the path of such a scenario, however, are two formidable levees: the levee on the Mississippi's west bank and the levee on the east bank of the Red River (see Figure 9.1). The two levee systems join about two miles south of Grand Cutoff Bayou. Bayou Cocodrie flows in the low ground between the two levees, along the base of the Mississippi levee's dry side before meandering across the intervening floodplain to join several smaller bayous and enter the Red River through an engineered gap in that river's levee (Figure 9.1).[31]

While the chances of a Mississippi River channel diversion here may be slight, no control structures are present to prevent it. And Widow Graham Bend has a troubling flood history. Despite the Corps' twenty-three-foot-high levee at Bougere, the levee crevassed in two places here during the 1927 flood.[32] In the event of a similar crevasse today, the Red River levee would impound the overflow temporarily, until the welling floodwaters overpower and widen the Bayou Cocodrie outlet to the Red River. Such a scenario is admittedly speculative; however, one cannot discount the possibility of a future Mississippi River flood culminating in a channel diversion at a spot other than Old River.

Time favors the Mississippi in the contest over the river's path to the Gulf of Mexico, whether the new channel begins at Old River or somewhere else. At the end of his book *Rising Tide*, about the 1927 flood, John M. Barry framed this contest in stark terms: "Keeping the Mississippi in its old channel has become by far the most serious engineering problem the Corps of Engineers faces."[33] The "problem" is our doing. Over the last three hundred years, we have casually sidled up to one of the world's largest rivers, perching cities on its banks and commandeering its floodplain for agriculture as if these were the shores of an eternal lake instead of the restless, twisting flanks of an alluvial giant.

Major General Charles C. Noble, who fought the 1973 flood, called the Lower Mississippi River the "greatest hydraulic challenge in the world."[34] Having accepted the challenge to oppose the river's natural process of building and abandoning its meander belts, we are committed to an escalating arms race demanding ever stronger control structures and higher levees. A little over fifty years have passed since the Corps stopped the Mississippi's channel shift into the Atchafalaya Basin. That alone is a significant victory over a river that routinely carries half a million cubic feet of water per second and can jump to more than twice that volume in a big flood. From the standpoint of geological time, however, the arms race has barely begun. We have to keep asking, "What will the levees and control structures look like a thousand years from now?"

From the days of Henry Shreve onward, Old River has been a place where important lessons have been taught, but not always learned, about the movement of water and sediment in a major river system. Unfortunately, well-intentioned human actions account for much of the challenge facing the Corps today. Shreve and *Heliopolis* proved that machines and dynamite could reshape the Mississippi's channel, sending the mighty current in a new direction. But Shreve's Turnbull Bend cutoff and the subsequent amputation of Raccourci Bend revealed the specter of chain reactions, unforeseen consequences that may take decades to unfold.

While a Mississippi River meander bend may be a navigational nuisance, the slow-forming curve holds a bundle of hydrologic variables in fragile equilibrium. After Shreve's cutoff, the removal of the tangled jam of trees that formed the Atchafalaya raft further disrupted the equilibrium at Old River. These two changes, seemingly as logical as they were shortsighted, helped the Mississippi evolve toward its future meander belt midway between the old Teche channel and the current one. The Corps of Engineers has prepared the Atchafalaya Basin for floodway purposes, but the Mississippi has a different destiny for this short, steep path to the Gulf.

Based on the approximate four-thousand-year life span of the Mississippi River's previous meander belt down the Bayou Teche route, the current meander belt's migration into the Atchafalaya Basin would be happening naturally without the help of human intervention. The river has no other place to go. American Indians, who were in the Mississippi Valley during the abandonment of the Teche channel and the establishment of the modern course, left no record of this spectacular event. Most likely, the people and their villages came through the change unscathed. Archaeological traces of hundreds of generations of river folk provide testimony that these people were well adapted to the Mississippi's wandering ways.[35] Their acceptance of the alluvial river in their midst allowed the Indians to thrive in its valley for around ten thousand years. By comparison, our three hundred years of levee-building and channel "improvements" don't bode well for a similar long-term relationship.

After the 1973 flood and the near destruction of the low sill, Louisiana State University professors Kazmann, Johnson, and Harris took the first responsible steps to examine the Mississippi's approaching diversion. Their suggestions for avoiding disaster call for bold changes in population settlement and infrastructure that, so far, have been ignored. As this book has endeavored to make clear, the problem is not going away. Without doubt, the US Army Corps of Engineers has done everything possible to keep the Mississippi in its familiar channel. Yet the main drawback to the Old River control structures is their permanence. Except for their moveable gates, the low-sill and auxiliary structures are fixed to the landscape while the river they were designed to control is dynamic and ever-changing. Perhaps future engineers will have the tools to build a control structure that can reconfigure its shape and move with the river. Until then, the Mississippi River will always hold the advantage in the contest.

In 1983, close on the heels of the Kazmann, Johnson, Harris report, engineer John D. Higby Jr completed a similar study of the Mississippi channel diversion for Colorado State University's Water Resources Institute titled *Possible Capture of the Mississippi by the Atchafalaya River*. Higby's Abstract or summary of his findings at the beginning of his report is succinct and worth quoting here:

> The possible diversion of the Mississippi River and man's effort to resist it, present one of the greatest river engineering problems ever encountered. The evidence that supports the claim that capture of the Mississippi by the Atchafalaya is forthcoming, is available and bountiful. Data on the deterioration of the capacity of the Mississippi below Old River and the increasing capacity of the Atchafalaya has been collected and authenticated. Neotectonic activity also indicates that the tendency toward diversion is increasing.[36]

Like Kazmann, Johnson, and Harris, Higby expected, given the weight of the evidence and the scope of the coming catastrophe, an official federal response to the diversion threat. Among his recommendations, he advocated the establishment of a congressional commission "made up of the world's foremost professionals in river engineering, geology, and water resources policy and planning" to examine all aspects of the problem and investigate viable solutions, adding that the commission's report should generate follow-up action. History has shown that, instead of acting, we have called the Mississippi's bluff.[37]

The most practical of Higby's suggestions, as bold as it is simplistic, is that we consider allowing the Mississippi to change its course at Old River.[38] Corps of Engineers geologist, Charles R. Kolb, also recognized the advantages of partnering with nature to permit the channel diversion to take place slowly under controlled conditions.[39] Such a paradigm shift away from battling the Mississippi River head-to-head in perpetuity is intimidating in scale and would require a national effort to equal our country's greatest achievements. In the Old River Control Complex, we have the mechanism in place for the control structures to effect a gradual and regulated channel diversion. More recently, other professionals have advocated this course of action, including University of Illinois ecologist Richard E. Sparks, who takes the approach that "the most sensible strategy is to work with the forces of nature rather than trying to overpower them."[40]

This alternative course of action makes sense but is fraught with troubling implications as well as exciting opportunities. Of foremost concern is the safe introduction of a new Mississippi River channel through the Atchafalaya Basin to the Gulf. The change in strategy would provide the Mississippi River Commission and the Corps of Engineers with the chance to design the Mississippi's path in order to minimize property damage and prevent loss of human lives. A major logistical challenge would be accommodating the Basin's bridges, trestles, and pipelines to the new waterway, but one might argue that this work is better done ahead of time instead of scrambling to repair things in the aftermath of a disaster. Likewise, with adequate planning, Morgan City need not be sacrificed to the channel diversion, which would be a likely consequence if the Mississippi overpowers the Old River structures at the height of a superflood. Protecting Morgan City would entail modification of the existing Atchafalaya River outlets to the Gulf. In the long run, the coastal towns of Morgan City, Franklin, and New Iberia could be positioned for a bright future with a new seaport developing in their midst.[41]

But what would happen to New Orleans? In the Wall Street parlance of the housing market crash of 2007, is the Crescent City too big to allow it to fail? Unlike other North American cities of comparable size, New Orleans stands exposed to three kinds of natural disaster: hurricanes, floods, and the Mississippi River's threatened diversion.[42] So far, the Corps of Engineers' floodways have offered adequate protection from the flood threat. The emergency outlets at Old River, Morganza, and Bonnet Carré should remain reliable until the coming of a flood greater than the hypothetical project flood. The city's vulnerability to hurricanes, however, is increasing with the disappearance of the coastal marshes between New Orleans and the open waters of the Gulf of Mexico.[43] In defense, multidisciplinary teams made up of ecologists, landscape architects, biologists, and others want to rebuild the marshes by diverting Mississippi River sediment out of the main channel into the coastal parishes.[44] Perhaps the Corps' bed-load experience at Old River can help with this initiative. If New Orleans can manage to dodge destruction from floods and storms, the channel diversion threat remains in the cards as a future disaster.

Allowing the Mississippi to divert the majority of its flow at Old River would remove the third threat—and most of the Crescent City's river as well. In Chapter 7, we looked at the difficulties facing New Orleans with a diminished Mississippi River flow. The city could not survive the prolonged exposure to saltwater that will invade the river channel. As a partial solution, Richard Sparks suggests the creation of two cities called New Orleans. In his scenario, Old New Orleans, the historic landmark and tourist destination, would remain much as it is, located on the high ground atop the ancient natural levee that attracted the first French colonists. Of course, Old New Orleans would require a source of fresh water, a logistical hurdle that may be too high. The new version of the city, destined to take over the Crescent City's role as an international port, would be built on the already developing coastal land near the mouth of the Atchafalaya River. Casting his vision far into the future, Sparks prophesies that the "new" New Orleans would have a finite life expectancy because of its location on the new Mississippi River. He estimates an Atchafalaya Basin residency of at least one thousand years before the Mississippi would again be in need of a new channel to the Gulf. At that time, the Corps of Engineers (or its future equivalent) could orchestrate the next channel diversion to send the great river back down its old path beside Lake Pontchartrain and reunite the commercial New Orleans with historic New Orleans.[45]

◆ ◆ ◆

NEW CONCEPTS like moveable cities and flexible control structures may be the only way for present-day US culture to coexist long-term with an active, meandering river as large as the Mississippi. Three hundred years after our forebears came to replace the Lower Mississippi Valley Indians, the expanding cities and farms crowding behind the levees expect the river to stay where it is, to remain an eternal, flowing lake. Below the latitude of Old River, that version of the Mississippi River is ending. The Corps of Engineers has done everything possible to maintain the status quo, but we now realize that "normal" cannot be a static river. Instead, the status quo is a river of change, responding to climatological and geological forces as well as to our interference.

For now, the modern mound builders who inherited the valley keep piling dirt and pouring concrete against the threat posed by a river confined unnaturally to a channel it needs to abandon. The levees and control structures help to create the illusion that the Mississippi River is a part of our permanent, natural landscape, much like the Grand Canyon. Tows push barges up and down the waterway, passing beneath bridges anchored to riverbanks that haven't changed in living memory. Tourists who visited New Orleans when they were children assume they will find the place much the same when they return with their grandchildren. Perhaps they will.

But the future holds a winter with heavy snows and rain from the Great Plains to the Ohio Valley, followed by springtime storm fronts bringing record rainfall to the same areas. As the Mississippi climbs the gage at Cairo, and the swollen river moves south, we need to remember Old River and the men and women of the Corps of Engineers who operate the control structures. They might be facing the map-changing flood no one thought would ever come.

Notes

Introduction

1. "Jazz Festival is Kicked Off," *Times-Picayune* (New Orleans), April 15, 1973; "Morganza To Open Tomorrow—Corps," *Times-Picayune* (New Orleans), April 16, 1973.

2. Roger T. Saucier, *Geomorphology and Quaternary Geologic History of the Lower Mississippi Valley.* Vol. 1 (Vicksburg, MS: US Army Corps of Engineers, prepared for the president, Mississippi River Commission, 1994), 263, Figures 25, 50; Elaine G. Yodis, Craig E. Colten, and David C. Johnson, *Geography of Louisiana* (Boston: McGraw-Hill, 2003), 24, Figure 2.8. Below Baton Rouge, the Mississippi's meander belt continued to periodically change its course, forming the modern outlet about one thousand years ago. Ibid., 38, Figure 2.19.

3. *Old River Control*, US Army Corps of Engineers, New Orleans District January 2009, 12.

4. David B. Johnson, "A Change in the Course of the Lower Mississippi River: Description and Analysis of Some Economic Consequences, Addendum B," in "If the Old River Control Structure Fails? The Physical and Economic Consequences," by Raphael G. Kazmann and David B. Johnson. *Louisiana Water Resources Research Institute*, Bulletin 12 (Baton Rouge: Louisiana State University, September 1980), 1; Kazmann and Johnson, "If the Old River Control Structure Fails?," 5, 22; *Old River Control*, 11; Raymond Harrison, Corps of Engineers, Old River lock, personal communication, April 17, 2015.

5. Kazmann and Johnson, "If the Old River Control Structure Fails?," 10–16.

6. Denise Grady, "Watch on the Mississippi," *Discover*, March 1983, 23–27; Kazmann and Johnson, "If the Old River Control Structure Fails?," 10–16, 21, 24–29, 41.

7. Richard Campanella, *Bienville's Dilemma: A Historical Geography of New Orleans* (Lafayette: Center for Louisiana Studies, University of Louisiana at Lafayette, 2008), 90–91; Calvin R. Fremling, *Immortal River: The Upper Mississippi*

in Ancient and Modern Times (Madison: University of Wisconsin Press, 2005), 12; Yodis et al., *Geography of Louisiana*, 27, Figure 2.13.

8. Andres Aslan, Whitney J. Autin, and Michael D. Blum, "Causes of River Avulsion: Insights from the Late Holocene Avulsion History of the Mississippi River, U.S.A.," *Journal of Sedimentary Research* 75 (2005): 650–664; Martin Reuss, *Designing the Bayous: The Control of Water in the Atchafalaya Basin, 1800–1995* (College Station: Texas A&M University Press, College Station, 2004), 243.

9. Throughout the Mississippi Valley, "Old River" is a common name for ox-bow lakes representing formerly active channels that became cut off from the main river.

10. Colin Thorne, Oliver Harmer, Chester Watson, Nick Clifford et al., *Current and Historical Sediment Loads in the Lower Mississippi River*, Final Report to United States Army, European Research Office of the US Army, Contract Number 1106-EN-01 (Nottingham, UK: School of Geography, University of Nottingham, University Park, Nottingham NG7 2RD, July 2008), iii.

11. Russell Beauvais, Old River Control, personal communication, September 8, 2015.

Language of the River

1. Charles A. Camillo, personal communication, March 26, 2015.

2. River Summary and Forecasts, Vicksburg District Corps of Engineers, http://155.76.244.230/riverstage/bullet.txt (accessed October, 6, 2015).

3. Vertical Datums, National Geodetic Survey, National Oceanic and Atmospheric, at Administration, http://www.ngs.noaa.gov/datums/vertical (accessed July, 7, 2015).

4. River Summary and Forecasts.

Chapter One

1. Reuss, *Designing the Bayous*, 28.

2. Loess, derived from the German verb *lösen*, meaning to loosen or dissolve, is pronounced "lurse." The term refers to the powdery soil's tendency to erode when stripped of covering vegetation. J. O. Snowdon and Richard R. Priddy, "Geology of Mississippi Loess," *Mississippi Geological, Economic, and Topographic Survey Bulletin* 111 (1968): 13–167. Loess deposits in the Lower Mississippi Valley date to the latter part of the Pleistocene Epoch, between 12,500 and 300,000 years ago.

Roger T. Saucier, *Geomorphology and Quaternary Geologic History of the Lower Mississippi Valley*, vol. 1, prepared for the president, US Army Corps of Engineers (Vicksburg, MS: Mississippi River Commission, 1994), 133–34.

3. William L. Murphy and Paul E. Albertson "Engineering Geological Geographical Information System of the Waterways Experiment Station," *Mississippi Geology* 17, no. 2 (June 1996): 36–37; C. D. Pearson and D. G. Hunter, "Moncla Gap and the Red River Diversion," in *Quaternary Geology and Geoarchaeology of the Lower Red River Valley: A Field Trip*, by Whitney J. Autin and Charles E. Pearson, Friends of the Pleistocene, South Central Cell, 11th Annual Field Conference, Alexandria, Louisiana, March 26–28, 1993, 99, Figure 8.2; Saucier, *Geomorphology and Quaternary Geologic History*, 215, 245, 246, Plates 2 and 28, Sheet 1.

4. Saucier, *Geomorphology and Quaternary Geologic History*, Plate 28, Sheets 2 and 3.

5. Yodis et al., *Geography of Louisiana*, 37–38, Figure 2.19.

6. Saucier, *Geomorphology and Quaternary Geologic History*, Figure 25.

7. Aslan et al., "Causes of River Avulsion," 659, Figure 11; Saucier, *Geomorphology and Quaternary Geologic History*, 247, 250–267, 281, Figures 25, 50, Plates 10, 11.

8. Saucier, *Geomorphology and Quaternary Geologic History*, 263.

9. Ibid., 317; Thorne et al. *Current and Historical Sediment Loads in the Lower Mississippi River*, iii, v, 58.

10. Aslan et al, "Causes of River Avulsion," Figure 12; Jeffrey A. Nittrouer, *Sediment Transport Dynamics in the Lower Mississippi River: Non-Uniform Flow and Its Effects on River-Channel Morphology*. (PhD diss., University of Texas at Austin, 2010), Figure 4.4, Table 2.2.

11. Karen Clauson, *Measuring Trends in Riverbed Gradation: A Lower Mississippi River Case Study* (master's thesis, Department of Geography and Environmental Resources, Southern Illinois University in Carbondale, May 2009), 9; Yodis et al., *Geography of Louisiana*, 21, 23.

12. Aslan et al., "Causes of River Avulsion," 661–63, Figure 13; Nittrouer, *Sediment Transport Dynamics in the Lower Mississippi River*, 23; Saucier, *Geomorphology and Quaternary Geologic History*, 263.

13. Aslan et al., "Causes of River Avulsion," 654–58, 662, Figures 6–9; Saucier, *Geomorphology and Quaternary Geologic History*, 9–11, Figure 4, Plate 28 Sheets 1 and 2; S. L. Perrault, C. E. Pearson, Carey L. Coxe, Sara A. Hahn III et al., *Archaeological Data Recovery at Angola Plantation, Sites 16WF121 and 16WF122 West Feliciana Parish, Louisiana*, Coastal Environments, Inc., Final Report prepared for New Orleans District US Army Corps of Engineers (2006), 13; E. J. Thomas, Report on Survey of Vicinity of Mouth of Red River, La., 1910–1911, with estimates of Cost of Closure Works, in Separation of Waters of Red River from Mississippi River. "River

Engineering Floods, 1896–1816," Technical Records, A-6 (formerly A-2), General
Collection, Mississippi River Commission, 24–26.

14. Saucier, *Geomorphology and Quaternary Geologic History*, 50, 53, 248, 250;
Yodis et al., *Geography of Louisiana*, 36–39, Figure 2.19.

15. Aslan et al., "Causes of River Avulsion," Figure 11.

16. Ibid.; Saucier, *Geomorphology and Quaternary Geologic History*, 267.

17. Aslan et al., "Causes of River Avulsion," Figure 11; Saucier, *Geomorphology
and Quaternary Geologic History*, Plate 1.

18. Ibid.; Pearson and Hunter, "Moncla Gap and the Red River Diversion,"
99–102; Saucier, *Geomorphology and Quaternary Geologic History*, 276.

19. Pearson and Hunter, "Moncla Gap and the Red River Diversion," 99–102;
Saucier, *Geomorphology and Quaternary Geologic History*, 276.

20. David G. Anderson and Kenneth E. Sassaman. "Modeling Paleoindian
and Early Archaic Settlement in the Southeast: A Historical Perspective," in *The
Paleoindian and Early Archaic Southeast*, ed. David G. Anderson and Kenneth E.
Sassaman (Tuscaloosa, AL: University of Alabama Press, 1996), Figure 2.1; Jeffrey
P. Brain, *Tunica Archaeology* (Cambridge, MA: Peabody Museum of Archaeology
and Ethnology, Harvard University, 1988), 94, 122, 124, 130, 134–36, 140, 145–47, 180;
Dennis Jones, *Cultural Resources Survey of Mile 306.3 to 293.4-R on the Mississippi
River, Concordia, Pointe Coupee and West Feliciana Parishes, Louisiana*, Cultural
Resources Series Report Number: COELMN/PD-91/103, Prepared for the US
Army Corps of Engineers, New Orleans District, Final Report, August 1993. (Baton
Rouge: Museum of Geoscience, Louisiana State University, 1993), 27–30, 53–65,
82–88; Samuel O. McGahey, *Mississippi Projectile Point Guide*, Archaeological
Report No. 31, rev. ed. (Jackson: Mississippi Department of Archives and History,
2004), 163–65.

21. Samuel O. McGahey, "A Compendium of Mississippi Dugout Canoes
Recorded since 1974," *Mississippi Archaeology* 21, no. 1 (1986): 58–70.

22. For details about the De Soto expedition, see Lawrence A. Clayton, Vernon
J. Knight Jr., and Edward C. Moore, *The De Soto Chronicles: The Expedition of
Hernando De Soto to North America in 1539–1543*, vols. 1, 2 (Tuscaloosa: University
of Alabama Press, 1993).

23. S. L. Perrault et al., *Archaeological Data Recovery at Angola Plantation*,
15; S. L. Perrault, Roger T. Saucier, Thurston H. G. Hahn III, Dayna Lee et al.,
*Cultural Resources Survey, Testing, and Exploratory Trenching for the Louisiana
State Penitentiary Levee Enlargement Project*, 10, *West Feliciana Parish, Louisiana*.
Baton Rouge, LA. Coastal Environments, Inc., 2001, http://www.dtic.mil/cgi bin/
GetTRDoc?AD=ADA387997&Location=U2&doc=GetTRDoc.pdf (accessed June
9, 2014).

24. Saucier, *Geomorphology and Quaternary Geologic History*, 109–12, 191–94, Figure 46.

25. Aslan et al., "Causes of River Avulsion," 662–663, Figures 6, 7, 11; Saucier, *Geomorphology and Quaternary Geologic History*, 105; Yodis et al., *Geography of Louisiana*, 31, Figure 2.16.

26. Charles M. Hudson, *Knights of Spain, Warriors of the Sun: Hernando de Soto and the South's Ancient Chiefdoms* (Athens: University of Georgia Press, 1997), 389–90; Saucier, *Geomorphology and Quaternary Geologic History*, 105–7, Figure 16.

27. Brain, *Tunica Archaeology*, 152, Figure 12.4; Richebourg G. McWilliams, trans. and ed., *Pierre Le Moyne d'Iberville, Iberville's Gulf Journals* (Tuscaloosa: University of Alabama Press, 1981), 69; Richebourg G. McWilliams, trans. and ed., *Fleur de Lys and Calumet: Being the Pénicaut Narrative of French Adventure in Louisiana* (Tuscaloosa: University of Alabama Press, 1953), 26; S. L. Perrault et al., *Cultural Resources Survey*, Figures 2–2, 3–13, 3–4.

28. *The Angola Story: Louisiana State Penitentiary, Angola, Louisiana, Burl Cain, Warden*, comp. by Louisiana State Penitentiary Museum Foundation, rev. ed. (West Felicia Parish: Louisiana State Penitentiary Museum, June 2011), 8.

29. Heather Mason, "Gator Guards Hunted at Angola," WAFB News, West Feliciana Parish, Louisiana, September 15, 2010, http://www.wafb.com/Global/story.asp?S=13162424&sms_ss=blogger (accessed June 22, 2013).

30. Swanton, *Indian Tribes of the Lower Mississippi Valley*, 285–86.

31. James F. Barnett Jr, *Mississippi's American Indians*, Heritage of Mississippi Series, vol. 6 (Jackson: University Press of Mississippi for the Mississippi Historical Society and the Mississippi Department of Archives and History, 2012), 104; Brain, *Tunica Archaeology*, 30–41, 152. Some French narratives indicate that the Houma village site was abandoned before the Tunicas arrived, while others indicate a violent expulsion. John R. Swanton, *Indian Tribes of the Lower Mississippi Valley and Adjacent Coast of the Gulf of Mexico*, Smithsonian Institution Bureau of American Ethnology *Bulletin* 43, 1911, repr. Dover Publications, 1998); Swanton, *Indian Tribes of the Lower Mississippi Valley*, 289. For information about the Indian slave trade, see Alan Gallay, *The Indian Slave Trade: The Rise of the English Empire in the American South, 1670–1717* (New Haven, CT.: Yale University Press, 2002).

32. Father Davion initially settled with the Tunicas in 1699 at the tribe's former location on the Lower Yazoo River, near present-day Vicksburg, Mississippi, and remained with them for some twenty years. The French named a Tunica Hills promontory landmark "La Roche Davion" (Davion's Rock) after the missionary. This bluff, near present-day Fort Adams, Mississippi, was later called Loftus Heights after a small force comprised of Tunicas and members of other local

tribes ambushed an English military expedition here led by Major Arthur Loftus (see Chapter 2). Barnett, *Mississippi's American Indians*, 104, 146–147; Brain, *Tunica Archaeology*, 152; Swanton, *Indian Tribes of the Lower Mississippi Valley*, 20, 289, 311–13.

33. McWilliams, trans. and ed., *Pierre Le Moyne d'Iberville, Iberville's Gulf Journals*, 122–23; Swanton, *Indian Tribes of the Lower Mississippi Valley*, 273.

34. Marc-Antoine Caillot, Erin M. Greenwald, ed., and Teri F. Chalmers, trans, *A Company Man: The Remarkable French-Atlantic Voyage of a Clerk for the Company of the Indies*, The Historic New Orleans Collection (2013), 84; McWilliams, *Pierre Le Moyne d'Iberville, Iberville's Gulf Journals*, 57.

35. Campanella, *Bienville's Dilemma*, 113–14; Shannon Lee Dawdy, *Building the Devil's Empire: French Colonial New Orleans*, vol. 1 (Chicago: University of Chicago Press, 2008), 193.

36. Craig E. Colton, *An Unnatural Metropolis: Wresting New Orleans from Nature* (Baton Rouge: Louisiana State University Press, 2005), 2; Ari Kelman, "Boundary Issues: Clarifying New Orleans's Murky Edges," *Journal of American History* 94 (December 2007): 695–703, http://www.journalofamericanhistory.org/projects/katrina/Kelman.html (accessed September 2, 2013; Pierce F. Lewis, *New Orleans: The Making of an Urban Landscape*, 2nd ed. (Santa Fe, NM: Center for American Places, 2003).

37. Dunbar Rowland and Albert G. Sanders, eds. and transls., *Mississippi Provincial Archives, 1729–1740, French Dominion*, vol. 3 (Jackson: Mississippi Department of Archives and History, 1927), 563.

38. Campanella, *Bienville's Dilemma*, 112; Rowland and Sanders, *Mississippi's Provincial Archives*, vol. 3, 343, 516.

39. Barnett, *Mississippi's American Indians*, 19–20; Saucier, *Geomorphology and Quaternary Geologic History*, 264.

40. Campanella, *Bienville's Dilemma*, 123, 125; Dawdy, *Building the Devil's Empire*, 15, 63–64, 81–82; Rowland and Sanders, *Mississippi's Provincial Archives*, vol. 2, 510; Ibid., vol. 3, 429, 563–64, 594, 637; Daniel H. Usner Jr., "From African Captivity to American Slavery: The Introduction of Black Laborers to Colonial Louisiana" in *The Louisiana Purchase Bicentennial Series in Louisiana History*, vol. 1: *The French Experience in Louisiana*, ed. Glenn R. Conrad (Lafayette: Center for Louisiana Studies, University of Southwestern Louisiana, 1995), 187–88; Samuel Wilson Jr. "Colonial Fortifications and Military Architecture in the Mississippi Valley," in *The Louisiana Purchase Bicentennial Series in Louisiana History*, vol. 1: *The French Experience in Louisiana*, ed. Glenn R. Conrad (Lafayette: Center for Louisiana Studies, University of Southwestern Louisiana, 1995), 384–85, 386–87.

41. Caillot, *A Company Man*, 78; Campanella, *Bienville's Dilemma*, 123; Antoine Simone Le Page du Pratz, *History of Louisiana or of the Western Parts of Virginia and Carolina* (1774; repr., Baton Rouge: Claitor's Publishing Division, 1972), 51; Kelman, "Boundary Issues."

Chapter Two

1. Edith McCall, *Conquering the Rivers: Henry Miller Shreve and the Navigation of America's Inland Waterways* (Baton Rouge: Louisiana State University Press, 1984), title page with portrait from the National Portrait Gallery, Smithsonian Institution, Washington DC.

2. Report of a majority of the Committee on the subject of the Raccourci Cut-Off, Louisiana Legislature 1846, 28; "Report of the Chief Engineer to the Secretary of War," *New York Spectator*, December 30, 1831; *Evening Post* (NY), August 3, 1831.

3. J. Parker Lamb, *Perfecting the American Steam Locomotive* (Bloomington: Indiana University Press, 2003), 1, 5.

4. Rodney A. Latimer and Charles W. Schweizer, *The Atchafalaya River Study: A report based upon engineering and geological studies of the enlargement of Old and Atchafalaya rivers including profiles and sections together with factual data which indicate the past rate and extent of progressive changes in Old and Atchafalaya rivers from their head through Grand and Six Mile lakes to the sea, all of which indicate the probable capture of the Mississippi River by the Atchafalaya River* (Vicksburg, MS: Corps of Engineers, Mississippi River Commission, 1951), Appendix B, 17; Christopher Morris, *The Big Muddy: An Environmental History of the Mississippi and Its Peoples from Hernando de Soto to Hurricane Katrina.* (Oxford: Oxford University Press, 2012), 96–97; Amos Stoddard, *Sketches of Louisiana, Historical and Descriptive* (1812; repr., Carlisle, MA: Applewood Books, n.d.), 167.

5. No longer noted on maps, Grand Point was probably located at the confluence of the Red River and Alligator (Long) Bayou, just above present-day Grand Bay (not to be confused with Grand Lake, located some six miles to the northwest). Grand Bay is a one-mile-long section of the Red River's former channel, approximately one mile north of the Red's pre-1831 confluence with the Mississippi. 1805 map by B. Lafton reproduced in Reuss, *Designing the Bayous*, 31; *Flood Control and Navigation Maps of the Mississippi River, Cairo, Illinois, to the Gulf of Mexico, Including Navigation Charts, Middle Mississippi River Below Hannibal, Missouri,* 28th ed. (Vicksburg, MS: Mississippi River Commission, Corps of Engineers, US Army, 1960), Map No. 41.

6. "Report made to the Legislature on the State of Louisiana," *Mississippi Statesman and Natchez Gazette*, May 31, 1827; Report of a majority of the Committee on the subject of the Raccourci Cut-Off, Louisiana Legislature 1846, 12, 13, 27, 36; Reuss, *Designing the Bayous*, 28, 30.

7. Dawdy, *Building the Devil's Empire*, 8.

8. Gerald O. Haffner, "Major Arthur Loftus' Journal of the Proceedings of His Majesty's Twenty-Second Regiment up the River Mississippi in 1764," *Journal of the Louisiana Historical Association* 20, no. 3 (Summer 1979): 325–34; Lieutenant Ross (Map 1775), "Course of the river Mississippi from Balise to Fort Chartres" (London: Robert Sayer, 1775), http://www.davidrumsey.com/luna/servlet/detail/ RUMSEY~8~1~3664~430011:Course-of-the-River-Mississipi,-fro (accessed June 16, 2014).

9. Brain, *Tunica Archaeology*, 39–42; George Gauld (Map 1778), "A Plan of the coast of part of West Florida & Louisiana: including the River Yazous. Surveyed by George Gauld M.A. for the Right Honourable the Board of Admiralty." Library of Congress, G4012.C6 1778 .G3 Vault, http://rla.unc.edu/EMAS/regions-ms. html#sec_f (accessed February 14, 2014); Ross (Map 1775), "Course of the river Mississippi from Balise to Fort Chartres."

10. Allen E. Begnaud, "Acadian Exile," *Journal of the Acadian Historical Association* 5, no. 1 (Winter 1964): 87–90; Carl A. Brasseaux, *The Founding of New Acadia: The Beginnings of Acadian Life in Louisiana, 1765–1803* (Baton Rouge: Louisiana State University Press, 1987), 5–74, Maps 2, 3.

11. Brian J. Costello, *A History of Pointe Coupée Parish, Louisiana* (Donaldsonville, LA: Margaret Media, 2010), 35–36; Gwendolyn Midlo Hall, *Africans in Colonial Louisiana: The Development of Afro-Creole Culture in the Eighteenth Century* (Baton Rouge: Louisiana State University Press, 1992), Figures 9, 10, 158, 227, 238, 277, 282, 285, 308, 302, 317; F. X. Charlevoix, Charles E. O'Neill, ed., and John Gilmary Shea, trans., *Charlevoix's Louisiana: Selections from the History and the Journal, Pierre F.X. de Charlevoix* (Baton Rouge: Louisiana State University Press, 1977), 163. (Note: Page numbers referenced for Charlevoix follow O'Neill's numbers at the bottom of each page in this edition.) Pointe Coupée (French for "cutoff point") marks the location of a natural Mississippi River cutoff that occurred in 1721. In 1700, French explorers found only about five hundred yards separating the converging meander loops and took advantage of the portage opportunity. The cutoff created the oxbow lake called False River at present-day New Roads, Louisiana. Flood Control and Navigation Maps of the Mississippi River (1960), Map 44; McWilliams, *Pierre Le Moyne d'Iberville, Iberville's Gulf Journals*, 66; McWilliams, *Fleur de Lys and Calumet*, 26.

12. Elias Durnford (Map 1771), "Plan of the River Mississippi from the River Yasous to the River Ibberville in West Florida. Showing lands granted and names

of proprietors," http://rla.unc.edu/EMAS/regions-ms.html#sec_f; George Gauld (Map 1778), "Plan of the coast of part of West Florida & Louisiana," http://rla.unc. edu/EMAS/regions-ms.html#sec_f; Walter Johnson, *Soul By Soul: Life Inside the Antebellum Slave Market* (Cambridge, MA: Harvard University Press, 1999), 5–6, 215; David J. Libby, *Slavery and Frontier Mississippi: 1720–1835* (Jackson: University Press of Mississippi, 2004), 19, 20, 22, 37; S. L. Perrault et al., *Archaeological Data Recovery at Angola Plantation, Sites 16WF121 and 16WF122 West Feliciana Parish, Louisiana*, 31; S. L. Perrault et al., *Cultural Resources Survey*, 28, 30–32, 40–57, Figures 3–9; William Wilton (Map 1774), "Part of the River Mississippi from Manchac up to the River Yazous" [2 sheets, 27.25" x 67.25".] [Mississippi River Commission (Vicksburg), 91.]

13. High Flows and Flood History on the Lower Mississippi River Below Red River Landing, LA (1543–present), Southern Regional Headquarters, National Oceanic and Atmospheric Administration, http://www.srh.noaa.gov/lix/?n=ms_flood_history; updated October 12, 2011 (accessed January 4, 2013).

14. Campanella, *Bienville's Dilemma*, 24.

15. Between 1826 and 1835, Francis Routh bought the land that eventually became Angola Penitentiary. Routh's partner, the notorious slave trader turned planter, Isaac Franklin, later acquired Routh's interest in the property. Wendell Holmes Stephenson, *Isaac Franklin: Slave Trader and Planter of the Old South, With Plantation Records* (1938; repr., Gloucester, MA: Peter Smith, 1968), 126–28. Franklin was married to Adelicia Hayes, later the mistress of Belmont Mansion in Nashville, Tennessee. Adelicia inherited the land at Franklin's death, and according to her biographer, it was she who named the riverside plantation Angola. Albert W. Wardin Jr., *Belmont Mansion: The Home of Joseph and Adelicia Acklen* (Nashville, TN: Belmont Mansion Association, 2005), 4.

16. Report of a majority of the Committee on the subject of the Raccourci Cut-Off, Louisiana Legislature 1846, 28; Reuss, *Designing the Bayous*, 31.

17. David F. Bastian, *Grant's Canal: The Union's Attempt to Bypass Vicksburg* (Shippensburg, PA: Burd Street Press, 1995); Carolyn Pace Davis, *The Winter of 1863: Grant's Louisiana Canal Expeditions*, Papers of the Blue and Gray Education Society, no. 4, BGES (Saline, MI: MacNaughton and Gunn, 1997); Danny W. Harrelson, "Geology of Grant's Canal: The Union's Attempt to Bypass Vicksburg, Mississippi," Paper No. 20-2, Southeastern Section—54th Annual Meeting, Geological Society of America, Biloxi, MS, March 18, 2005.

18. Shreve's report of this operation appeared in the December 30, 1831, edition of the *New York Spectator*. In the report, he refers to this cutoff as Burch's Bend; however, the cutoff is labeled Bunch's Bend on present-day maps of the area.

19. H. B. Ferguson, *History of the Improvement of the Lower Mississippi River for Flood Control and Navigation, 1932–1939* (Vicksburg, MS: US Army Corps

of Engineers, Mississippi River Commission, 1940), Plate 11; *Flood Control and Navigation Maps of the Mississippi River, Cairo, Illinois, to the Gulf of Mexico*, Map 29; *Evening Post* (NY), August 3, 1831.

20. McCall, *Conquering the Rivers*, 16, 122–26, 130–31, 136–47, 149–56.

21. Report of a majority of the Committee on the subject of the Raccourci Cut-Off, Louisiana Legislature 1846, 28.

22. "Navigation: Ohio and Mississippi Rivers," *Natchez Gazette* (MS), May 29, 1830.

23. McCall, *Conquering the Rivers*, 187–88; Technology: 1828–1840, Cleaning up the Mississippi, Illinois State Museum, http://www.museum.state.il.us/RiverWeb/landings/Ambot/TECH/TECH12.htm (accessed April 26, 2014).

24. McCall, *Conquering the Rivers*, 184–85; Karen M. O'Neill, *Rivers by Design: State Power and the Origins of U.S. Flood Control* (Durham, NC: Duke University Press, 2006), 41.

25. "Report of the Chief Engineer to the Secretary of War," *New York Spectator*, December 30, 1831.

26. "The Mississippi," *New York Herald*, July 7, 1830.

27. Jones, *Cultural Resources Survey of Mile 306.3 to 293.4-R on the Mississippi River*, Table 1.

28. "Report of the Chief Engineer to the Secretary of War," *New York Spectator*, December 30, 1831; *Evening Post* (New York), August 3, 1831.

29. Fielding Lucas (Map 1823), "Mississippi" [Baltimore], Mississippi Department of Archives and History, Archives Library Collection MA/92.0062a, Jackson; Henry S. Tanner (Map 1820), "Louisiana and Mississippi" [Philadelphia], Mississippi Department of Archives and History, Jackson, MA/87.0006(c).

30. Based upon contemporary sources, Shreve's cut was 17 feet wide and 22 feet deep and ran a distance of 160 feet. In discussing Shreve's cutoff and the Raccourci cutoff, I use the estimation that one cubic yard of earth weighs approximately one ton. Report of a majority of the Committee on the subject of the Raccourci Cut-Off, Louisiana Legislature 1846, 28; *Evening Post* (NY), August 3, 1831.

31. Report of a majority of the Committee on the subject of the Raccourci Cut-Off, Louisiana Legislature 1846, 28; *Evening Post* (NY), August 3, 1831.

32. S. L. Perrault et al., *Archaeological Data Recovery at Angola Plantation, Sites 16WF121 and 16WF122 West Feliciana Parish, Louisiana*, 15. See the discussion about point bar formation in Chapter 1 and Saucier, *Geomorphology and Quaternary Geologic History of the Lower Mississippi Valley*, 109–12, 191–94, Figure 46, Plates 10–11.

33. "Report of the Chief Engineer to the Secretary of War," *New York Spectator*, December 30, 1831; *Evening Post* (New York), August 3, 1831.

34. Report of a majority of the Committee on the subject of the Raccourci Cut-Off, Louisiana Legislature 1846, 8, 10, 11.

35. Ibid., 4, 11.

36. Charles A. Camillo, *Divine Providence: The 2011 Flood in the Mississippi River and Tributaries Project* (Vicksburg: Mississippi River Commission, 2012), 10; "From the Army and Navy Chronicle," *New York Spectator*, November 15, 1838; Report of a majority of the Committee on the subject of the Raccourci Cut-Off, Louisiana Legislature 1846, 12, 20, 24, 36.

37. Reuss, *Designing the Bayous*, 19–29.

38. Kermit L. Hebert, "The Flood Control Capabilities of the Atchafalaya Basin Floodway," *Louisiana Water Resources Research Institute, Bulletin GT-1*, April 1967, Louisiana State University, Baton Rouge, 12, 40; Kazmann and Johnson, "If the Old River Control Structure Fails," 7; Reuss, *Designing the Bayous*, 28–36.

39. "Red River Raft," *Evening Post* (NY), May 23, 1838; Ibid., July 27, 1838; Ibid., June 6, 1839; Ibid., June 9, 1845; Annual Report of P. O. Hébert, State Engineer to the [Louisiana] Legislature, January 1847 (New Orleans: Magne & Weisse, 1847), 3–5; McCall, *Conquering the Rivers*, 190–218; Mitchel Whittington, *No Hope! The Story of the Great Red River Raft* (St. Francisville, LA: House Publishing, 2009).

40. The Attakapas (also spelled "Atakapa") constituted an American Indian tribal group located in present-day southwest Louisiana and along the southeast Texas coastal area in the eighteenth century. Swanton, *Indian Tribes of the Lower Mississippi Valley*, 360–63.

41. Gay M. Gomez, "Describing Louisiana: The Contribution of William Darby," *Journal of the Louisiana Historical Association* 34, no. 1 (Winter 1993): 102; Martin Reuss, "The Army Corps of Engineers and Flood-Control Politics on the Lower Mississippi," *Journal of the Louisiana Historical Association* 23, no. 2 (Spring 1982): 137.

42. Reuss, *Designing the Bayous*, 30.

43. Ibid., 28–33; Jones, *Cultural Resources Survey of Mile 306.3 to 293.4-R on the Mississippi River*, 16.

44. "A New Notion," *Times-Picayune* (New Orleans), December 25, 1842.

45. Annual Report of P. O. Hébert, State Engineer to the [Louisiana] Legislature, January 1847, 7.

46. Report of a majority of the Committee on the subject of the Raccourci Cut-Off, Louisiana Legislature 1846, 7.

47. Annual Report of P. O. Hébert, State Engineer to the [Louisiana] Legislature, January 1847, 6–7; Joseph Holt Ingraham, *The Southwest by a Yankee* (1835; repr., Readex Microprint Corp., 1966), vol. 2, 104–5, 171, 190; Report of a majority of the Committee on the subject of the Raccourci Cut-Off, Louisiana Legislature 1846, 5–7, 9–11, 20–21, 25.

48. "From the Army and Navy Chronicle," *New York Spectator*, November 15, 1838; "Mississippi and Red River," *Evening Post* (NY), October 17, 1838.

49. Gauld (Map 1778), "A Plan of the coast of part of West Florida & Louisiana."

50. The Ofos (sometimes called Ofogoulas) constituted a small tribal group allied with the Tunicas. Jeffrey P. Brain, George Roth, and Willem J. De Reuse, "Tunica, Biloxi, and Ofo," in *Handbook of North American Indians*, vol. 14: *Southeast*, gen. ed. William C. Sturtevant, vol. ed. Raymond D. Fogelson (Washington, DC: Smithsonian Institution, 2004), 594–97.

51. Nineteenth-century estimates vary on the distance around Raccourci Bend. Some documents generated by the Louisiana legislature in 1846 report the distance around the bend to be twenty-nine miles. Report of a majority of the Committee on the subject of the Raccourci Cut-Off, Louisiana Legislature 1846, 3. Nineteenth-century and present-day maps indicate that an estimate of nineteen miles is more accurate.

52. Report of a majority of the Committee on the subject of the Raccourci Cut-Off, Louisiana Legislature 1846, 1–14.

53. Ibid., 23–30.

54. Ibid., 16–17.

55. Acts Passed at the Second Session of the Sixteenth Legislature of the State of Louisiana, 1844, 60–61; Report of a majority of the Committee on the subject of the Raccourci Cut-Off, Louisiana Legislature 1846, 17; *Times-Picayune* (New Orleans), March 12, 1843.

56. Mark T. Carleton, *Politics and Punishment: The History of the Louisiana State Penal System* (Baton Rouge: Louisiana State University Press, 1971); S. L. Perrault et al., *Archaeological Data Recovery at Angola Plantation, Sites 16WF121 and 16WF122 West Feliciana Parish, Louisiana*, 93.

57. "Raccourci Cut-Off," *New Orleans Commercial Bulletin*, November 20, 1844; Ibid., December 24, 1844.

58. Thomas W. Cutrer, "Hébert, Paul Octave," *Handbook of Texas Online*, http://www.tshaonline.org/handbook/online/articles/fhe09 (accessed July 20, 2014).

59. Annual Report of the State Engineer to the Legislature of the State of Louisiana, February 1846 (New Orleans: Printed at the Jeffersonian Office, 1846), 8; "Raccourci Cut-Off," *New Orleans Commercial Bulletin*, October 30, 1844.

60. Annual Report of the State Engineer to the Legislature of the State of Louisiana, February 1846, 3–9.

61. Ibid., 9–10.

62. Ibid., 10–11.

63. Ibid.

64. Report of a majority of the Committee on the subject of the Raccourci Cut-Off, Louisiana Legislature 1846, 1–14.

65. Ibid., 14–37.

66. "Louisiana Legislature," *Times-Picayune* (New Orleans), May 17, 1846.

67. Annual Report of P. O. Hébert, State Engineer to the [Louisiana] Legislature, January 1847, 1–7.

68. Ibid., 1–6; Maps Showing the Progressive Development of US Railroads, 1830–1950, http://www.cprr.org/Museum/RR_Development.html#1L (accessed July 25, 2014).

69. Acts Passed at the Second Session of the First Legislature of the State of Louisiana, Held and Begun in the City of New Orleans, January 11, 1847 (New Orleans: State Printer, 1847), 58–59.

70. Jones, *Cultural Resources Survey of Mile 306.3 to 293.4-R on the Mississippi River*, 51; Journal of the Senate of the State of Louisiana, Session of 1848 (New Orleans: Printed at the Office of the *Louisiana Courier*), 93.

71. Mark Twain, *Life on the Mississippi* (1883; repr., Mineola, NY: Dover Publications, 2000), 89.

72. Norman's Chart of the Lower Mississippi River by A. Persac (New Orleans: B. M. Norman, 1858); J. C. Rowland and W. E. Dietrich, "The Evolution of a Tie Channel," in *River, Coastal and Estuarine Morphodynamics: RCM 205*, ed. G. Parker and M. Garcia (London: Taylor and Francis, 2006), Figure 1, 733, Department of Earth & Planetary Science, University of California-Berkeley, 2006, http://eps.berkeley.edu/~bill/papers/rowtet_135.pdf (accessed July 26, 2014).

73. B. M. Harrod, "The Levees of Louisiana: Their Condition and Requirements," *Times-Picayune* (New Orleans), December 6, 1878; "The Bayou Plaquemine," *Times-Picayune* (New Orleans), March 7, 1858.

74. Acts Passed at the Second Session of the Sixteenth Legislature of the State of Louisiana, 1844, 60–61; "History," Louisiana's Old State Capitol, http://louisianaoldstatecapitol.org/PageDisplay.asp?p1=805 (accessed July 28, 2014); "Internal Improvements," *Times-Picayune* (New Orleans), January 19, 1848; "Louisiana Legislature," *Times-Picayune* (New Orleans), March 9, 1853; "The Outlets of the Mississippi," *Times-Picayune* (New Orleans), January 21, 1859; "The Bayou Plaquemine," *Times-Picayune* (New Orleans), March 7, 1858.

75. Reuss, *Designing the Bayous*, 33, 73, 209; "The Atchafalaya," *Times-Picayune* (New Orleans), August 23, 1863.

76. A. A. Humphreys and H. L. Abbott, Report upon the Physics and Hydraulics of the Mississippi River; upon the Protection of the Alluvial Region Against Overflowing; and upon the Deepening of the Mouths: Based upon surveys and investigations made under the Acts of Congress directing the topographical and

hydrographical survey of the delta of the Mississippi River, with such investigations as might lead to determine the most practicable plan for securing it from
inundation, and the best mode of deepening the channels of the mouths of the
river. Submitted by the Bureau of Topographical Engineers, War Department 1861
(Washington, DC: Government Printing Office, 1867), 125; Amos Stickney, Red and
Atchafalaya Rivers, Progress of Surveys and Examinations, in Annual Report of
the Chief of Engineers, United States Army, to the Secretary of War, for the Year
1884, Part 4 (Washington, DC: Government Printing Office, 1926), 2420–21.

    77. Reuss, *Designing the Bayous*, 26–28, 73.

    78. "A River Seceding," *New York Herald*, March 26, 1861; Stickney, Red and
Atchafalaya Rivers, Progress of Surveys and Examinations, 2420–21.

Chapter Three

    1. "Old River and the Mud Hole," *Times-Picayune* (New Orleans), September
18, 1877.

    2. Annual Report of the State Engineer to the Legislature of the State of
Louisiana, February 1846, 8–9; Reuss, *Designing the Bayous*, 210; "The Bayou
Plaquemine," *Times-Picayune* (New Orleans), March 7, 1858; "The Outlets of the
Mississippi," *Times-Picayune* (New Orleans), January 21, 1859.

    3. Campanella, *Bienville's Dilemma*, 28, 34, 205, 309; High Flows and Flood
History on the Lower Mississippi River; Humphreys and Abbott, Report upon the
Physics and Hydraulics of the Mississippi River, 22–23, 116.

    4. Campanella, *Bienville's Dilemma*, 205; High Flows and Flood History on the
Lower Mississippi River; Reuss, *Designing the Bayous*, 57, 62–63.

    5. Humphreys and Abbott, Report upon the Physics and Hydraulics of the
Mississippi River, 150–51, 157–92; Jamie Moore and Dorothy P. Moore, *The Army
Corps of Engineers and the Evolution of Federal Flood Plain Management Policy*,
Program on Environment and Behavior, Special Publication No. 20 (Boulder, CO:
Institute of Behavioral Science, University of Colorado, 1989), 1–2.

    6. Humphreys and Abbott, Report upon the Physics and Hydraulics of the
Mississippi River, 150–58, 162–76.

    7. Map of the Cairo & Fulton Railroad Exhibiting the principal tributary lines
as projected and its connection with other Railroads west of the Mississippi River,
which unite with the Missouri Pacific Railroad and the south projected Pacific
Railroad via El Paso to the Pacific Ocean, showing also the Connection by Railroad of the City of New Orleans & St. Louis, comp. and drawn by I. Wilamowicz,
Little Rock, AR, September 1853, http://usgwarchives.org/maps/louisiana/state-
map/la1853.jpg (accessed August 18, 2014).

8. John M. Barry, *Rising Tide: The Great Mississippi Flood of 1927 and How It Changed America* (New York: Simon and Schuster, 1997), 62, 71–79, 83–85, 87, 90–91; Humphreys and Abbott, Report upon the Physics and Hydraulics of the Mississippi River, 187. The Italian hydraulic engineer Domenico Guglielmini, active in the late-seventeenth and early-eighteenth centuries, pioneered the approach that Eads followed in confining a river channel to concentrate the current and scour the bed. Hebert, "Flood Control Capabilities of the Atchafalaya Basin Floodway," 35–36.

9. Humphreys and Abbott, Report upon the Physics and Hydraulics of the Mississippi River, 157–158; Moore and Moore, *Army Corps of Engineers and the Evolution of Federal Flood Plain Management Policy*, 2; Reuss, *Designing the Bayous*, 64–66.

10. John Ewens, Red River Landing and Head of the Atchafalaya, in Annual Report of the Chief of Engineers, United States Army, to the Secretary of War, for the Year 1884, Part IV, Washington, DC, 2636, 2639.

11. Jones, *Cultural Resources Survey of Mile 306.3 to 293.4-R on the Mississippi River*, 45; "Trip on the Atchafalaya River—Drawn by J. O. Davidson," *Harper's Weekly*, April 14, 1883, 237; Julian Oliver Davidson, Artist, http://www.battleoflakeerieart.com/jodartist.php (accessed August 16, 2014). The 1858 Persac map of the Mississippi River shows a hotel and store at the Red River Landing. Norman's Chart of the Lower Mississippi River by A. Persac, 1858.

12. Executive Documents of the House of Representatives, Third Session of the Forty-Sixth Congress: 1880-1881, vol. 3, Engineers, no. 1, part 2 (Washington, DC: Government Printing Office, 1881); Latimer and Schweizer, *Atchafalaya River Study*, Appendix D, 10.

13. Ewens, Red River Landing and Head of the Atchafalaya, 2636, 2638–39.

14. Ibid.

15. Report of the Board of State Engineers. State of Louisiana, Office Board of Engineers, New Orleans, LA, April 21, 1884, 115; Ibid., April 20, 1896, 7; Riparian Damages on the East Bank of the Mississippi River, Hearings before the United States Congress, House Committee on the Judiciary, House of Representatives, 62nd Cong., 2d sess. on H.R. 19412, March 22, 1912 (Washington, DC: Government Printing Office, 1912), 55–56.

16. Ewens, Red River Landing and Head of the Atchafalaya, 2639.

17. Ibid., 2640.

18. Ibid.

19. Ibid., 2640–41.

20. Ibid.

21. Annual Report of the Chief of Engineers, 1888, Part IV, 2296, 2308–9.

22. Ibid., 2309–10; Ibid., 1889, Part IV, 2723–24, 2739.

23. Ibid., 1885, Part IV, 2825; Ibid., 1888, Part IV, 2298–99; "History of the Prison," Angola Museum, Louisiana State Penitentiary Museum Foundation 2013, http://angolamuseum.org/?q=History#history (accessed March 31, 2013).

24. Annual Report of the Chief of Engineers, 1889, Part IV, 2726–2727.

25. Ibid., 1885, Part IV, 2825; Ibid., 2296; Ibid., 1889, Part IV, 2724.

26. Ibid., 1888, Part IV, 2296, 2298, 2308; Ibid., 1889, Part IV, 2723, 2724.

27. Ibid., 1888, Part IV, 2300.

28. Ibid., 1888, Part IV, 2299; Past MRC [Mississippi River Commission] Members, US Army Corps of Engineers, Mississippi Valley Division, http://www.mvd.usace.army.mil/About/MississippiRiverCommission(MRC)/PastMRCMembers.aspx (accessed August 28, 2014).

29. Annual Report of the Chief of Engineers, 1885, Part IV, 2827; Ibid., 1888, Part IV, 2299–2300; Ibid., 1889, Part IV, 2726–2727. Stickney's plan is much the same as a plan developed in 1874 by Louisiana assistant engineer E. H. Angemar, Annual Report of the Secretary of War, vol. 2, Part 2 (Washington, DC: Government Printing Office, 1880), 1288–89.

30. Annual Report of the Chief of Engineers, 1885, Part IV, 2560–69; Ibid., 1888, Part IV, 2300; Ibid., 1889, Part IV, 2724–41.

31. Davis R. Dewey, *Financial History of the United States*, 6th ed. (New York: Longmans, Green, 1918), 440–69; Report of the Chief of Engineers, US Army, 1911, Part III, 3179; Reuss, *Designing the Bayous*, 81–83.

32. Reuss, *Designing the Bayous*, 34, 53, 90; Mark T. Swanson, Interpretive Program, Plaquemine Locks State Commemorative Area, Project No. 06-06-00-78-12, New World Research, Inc., Report of Investigations No. 82-30, State of Louisiana, Office of State Parks, November 28, 1983, 61, 62. In 1961, the bigger lock at Port Allen near Baton Rouge replaced the Plaquemine Lock and became a component of Louisiana's intracoastal waterway; Reuss, *Designing the Bayous*, 237.

33. Annual Report of the Chief of Engineers, 1885, Part IV, 2555, 1559, 2827; High Flows and Flood History on the Lower Mississippi River; Reuss, *Designing the Bayous*, 89; Stickney, Red and Atchafalaya Rivers, Progress of Surveys and Examinations, 2419–21.

34. A. E. Morgan et al., *A Preliminary Report on the Drainage of the Fifth Louisiana Levee District, Comprising the Parishes of East Carroll, Madison, Tensas, and Concordia*, US Department of Agriculture, Office of Experiment Stations—Circular 104 (Washington, DC: Government Printing Office, 1911), 3, 5, 19–20; Annex R, Red River Backwater Area, Mississippi River and Tributaries Comprehensive Review Report, vol. 5 in Mississippi River and Tributaries Project: Letter from the Secretary of the Army transmitting a letter from the Chief of Engineers, Department of the Army, dated April 6, 1963, submitting a report, together with accompanying papers and illustrations, on a review of the Mississippi River and

Tributaries Project, in response to a resolution adopted June 12, 1964, by the Committee on Public Works of the US Senate (Washington, DC: Government Printing Office, 1964), 2.

35. High Flows and Flood History on the Lower Mississippi River; Morgan et al., *A Preliminary Report on the Drainage of the Fifth Louisiana Levee District*, 20–35.

36. E. J. Thomas, letter to Charles L. Potter, USACE and Secretary, Mississippi River Commission, May 15, 1912, in Separation of Waters of Red River from Mississippi River, "River Engineering Floods, 1896–1816," Technical Records, A-6 (formerly A-2), General Collection, Mississippi River Commission.

37. Divorcement of Mississippi and Atchafalaya Rivers, Brief of Atchafalaya Protective Association, submitted to the Mississippi River Commission. 1896–1816, Technical Records, A-2, General Collection, Mississippi River Commission, 2, 3, 4, 6, 10, 17, 21, 22, 31, 33.

38. Ibid., 6, 13, 16.

39. Annual Report of the Chief of Engineers, 1885, Part IV, 2554.

40. Divorcement of Mississippi and Atchafalaya River, Brief of Atchafalaya Protective Association.

41. J. F. Coleman, Consulting Engineer, Atchafalaya Protective Association, letter to the Mississippi River Commission, April 20, 1911, in Separation of Waters of Red River from Mississippi River, "River Engineering Floods, 1896–1816," Technical Records, A-6 (formerly A-2), General Collection, Mississippi River Commission.

42. High Flows and Flood History on the Lower Mississippi River.

43. Separation of Waters of Red River from Mississippi River, "River Engineering Floods, 1896–1816."

44. Ibid.

45. Ibid.

46. Thomas, Report on Survey of Vicinity of Mouth of Red River, LA, 1910–1911, 1–6.

47. Ibid., 7–9, 24–26.

48. Ibid., 10–11.

49. High Flows and Flood History on the Lower Mississippi River; Separation of Red and Atchafalaya Rivers from Mississippi River, 2–7.

50. Separation of Red and Atchafalaya Rivers from Mississippi River, 2–7.

51. High Flows and Flood History on the Lower Mississippi River.

52. *John F. Cubbins, Appt. v. Mississippi River Commission and the Yazoo-Mississippi Delta Levee Board, Supreme Court Reporter*, vol. 36, Cases Argued and Determined in the United States Supreme Court, October Term, 1915; December, 1915-July, 1916; (St. Paul, MN: West Publishing Company, 1916), 676.

53. Barry, *Rising Tide*, 79; Humphreys and Abbott, Report upon the Physics and Hydraulics of the Mississippi River, 176–192.

54. Matthew D. Pearcy, "A History of the Randsdell-Humphreys Flood Control Act of 1917," *Journal of the Louisiana Historical Association* 41, no. 2 (2000): 135, 155–56; Reuss, *Designing the Bayous*, 93–96.

55. High Flows and Flood History on the Lower Mississippi River.

56. Thomas, Report on Survey of Vicinity of Mouth of Red River, LA, 1910–11, 3–4, 6.

57. Moore and Moore, *Army Corps of Engineers and the Evolution of Federal Flood Plain Management Policy*, 6.

58. Report of the Chief of Engineers, US Army, 1926, in Two Parts, Part I, 1818.

Chapter Four

1. Estimates vary on the volume of the 1927 floods. This estimate is from Brigadier General Harley B. Ferguson, who served as president of the Mississippi River Commission from 1932 to 1939. Hearings before the Committee on Commerce, United States Senate, 75th Cong., 3d sess. on S. 3354, a bill to amend the act entitled "An Act for the Control of Floods on the Mississippi River and its Tributaries, and for other purposes," approved May 15, 1928; March 28, 29, and 30, 1938, printed for the use of the Committee on Commerce (Washington, DC: Government Printing Office, 1938), 52–53. John M. Barry has noted that some engineers estimate the maximum volume of the 1927 floods to have been as high as 3 million cubic feet per second. Review of *Beyond Control* manuscript for University Press of Mississippi, February 2016. The average rate of discharge on the Lower Mississippi River is around 600,000 cubic feet per second, US EPA Office of Wetlands, Oceans, and Watersheds.

2. Jim Bradshaw, "Great Flood of 1927," in *Know LA: Encyclopedia of Louisiana*, ed. David Johnson, Louisiana Endowment for the Humanities, 2010, article published May 13, 2011, http://www.knowla.org/entry/763 (accessed October 27, 2014).

3. Camillo, *Divine Providence*, 128–29; Comprehensive Report on Reservoirs in Mississippi River Basin: Letter from the Secretary of War transmitting pursuant to Section 10 of the Flood Control Act approved May 15, 1928, a letter from the Chief of Engineers, United States Army, dated July 26, 1935, submitting a report, together with accompanying papers and illustrations, on further flood control of the Lower Mississippi River by control of floodwaters in the drainage basins of the tributaries by establishment of a reservoir system (Washington, DC: Government Printing Office, 1936), 76–77; Hearings before the Committee on Commerce, United States Senate (1938), 52–53; "Jonesville Hit Hard By Flood," *Concordia Sentinel* (LA), July 9, 1927.

4. Barry, *Rising Tide*, 283–85; Bradshaw, "Great Flood of 1927"; D. O. Elliott, *The Improvement of the Lower Mississippi River for Flood Control and Navigation*, Plate XXXVII (Vicksburg, MS: US Waterways, 1952).

5. Hearings before the Committee on Flood Control, House of Representatives (1933), 73rd Cong., 1st sess., March 30, May 12, 1933 (Washington, DC: Government Printing Office, 1933), 21.

6. Barry, *Rising Tide*, 227–58; Bradshaw, "Great Flood of 1927"; Reuss, *Designing the Bayous*, 98–99, 104–5.

7. Barry, *Rising Tide*, 255–257; Camillo, *Divine Providence*, 128–33; Elliott, *Improvement of the Lower Mississippi River for Flood Control and Navigation*, Plate XXXVII; O'Neill, *Rivers by Design*, 138; Reuss, *Designing the Bayous*, 100–101, 103.

8. Flood Control in the Mississippi Valley, 70th Cong., 1st sess., House of Representatives, Document No. 90, December 8, 1927, 3–7.

9. Ibid., 1.

10. "Edgar Jadwin, Major General, United States Army," Arlington National Cemetery, http://www.arlingtoncemetery.net/ejadwin.htm (accessed December 14, 2014); Flood Control Act of 1928, Section1, 70th Cong., sess. 1, chap. 596, May 15, 1928, http://www.mvd.usace.army.mil/Portals/52/docs/MRC/Appendix_E._1928_Flood_Control_Act.pdf (accessed September 25, 2014); Flood Control in the Mississippi Valley, 3–4; Humphreys and Abbott, Report upon the Physics and Hydraulics of the Mississippi River, 102, 110, 143, 144, 164, 166, 168, 164, 171; *Flood Control and Navigation Maps of the Mississippi River, Cairo, Illinois, to the Gulf of Mexico*, Map No. 50; Annual Report of the Mississippi River Commission for the Fiscal Year ending June 30, 1894, maps of crevasses follow the Mississippi River Commission report; Reuss, *Designing the Bayous*, 111; Norman's Chart of the Lower Mississippi River by A. Persac (1858).

11. Campanella, *Bienville's Dilemma*, 21, 195; Humphreys and Abbott, Report upon the Physics and Hydraulics of the Mississippi River, 30, 52, 55, 57, 164, 171; Norman's Chart of the Lower Mississippi River by A. Persac (1858).

12. Bonnet Carré's Spillway Overview; Flood Control in the Mississippi Valley, 27; Humphreys and Abbott, Report upon the Physics and Hydraulics of the Mississippi River, 171; O'Neill, *Rivers by Design*, 144.

13. Flood Control in the Mississippi Valley, 6–29; O'Neill, *Rivers by Design*, 144; The Mississippi River & Tributaries Project: Birds Point-New Madrid Floodway, 3–7.

14. Barry, *Rising Tide*, 160; Camillo, *Divine Providence*, 130; Harrison and Kellmorgan, "Socio-Economic History of Cypress Creek Drainage District," 21, 32; O'Neill, *Rivers by Design*, 132.

15. Camillo, *Divine Providence*, 130–32; Flood Control in the Mississippi Valley, 6–7, 27–28; Matthew Reonas, "Delta Planters and the Eudora Floodway:

The Politics of Persistence in 1930s Louisiana," *Journal of the Louisiana Historical Association* 50, no. 2 (Spring 2009): 162–64.

16. Camillo, *Divine Providence*, 128–29; Flood Control in the Mississippi Valley, 6.

17. Flood Control in the Mississippi Valley, 24–26, 31; Hearings before the Committee on Commerce, United States Senate (1938), 105.

18. Camillo, *Divine Providence*, 10, 132; Flood Control in the Mississippi Valley, 26.

19. Camillo, *Divine Providence*, 8–14, 117–23, 248–49; Flood Control in the Mississippi Valley, 24; Reuss, *Designing the Bayous*, 113.

20. Camillo, *Divine Providence*, 10; Comprehensive Report on Reservoirs in Mississippi River Basin, 83–89; Flood Control in the Mississippi Valley, 24; Hearings before the Committee on Commerce, United States Senate (1938), 52–53; Hebert, "Flood Control Capabilities of the Atchafalaya Basin Floodway," 38; High Flows and Flood History on the Lower Mississippi River; The Mississippi River and Tributaries Project: Designing the Project Flood.

21. Camillo, *Divine Providence*, 130–32; "Flood Control Ass'n [*sic*] Adopts Resolution," *Concordia Sentinel* (LA), December 12, 1927; Flood Control in the Lower Mississippi Valley: Report Submitted by the Board of State Engineers, 22; Flood Control in the Mississippi Valley, 20, 31; Reuss, *Designing the Bayous*, 113–17.

22. Flood Control Act of 1928, Section 1; Flood Control in the Mississippi Valley, 33–34.

23. Camillo, *Divine Providence*, 132; Flood Control in the Mississippi Valley, 8; Special Report of the Mississippi River Commission, v; Mississippi River & Tributaries Project: Birds Point-New Madrid Floodway, 4.

24. Camillo, *Divine Providence*, 133–34; Flood Control in the Mississippi Valley, 8, 9, 11, 14, 34; Pearcy, "After the Flood: A History of the 1928 Flood Control Act," 181–82, 187.

25. Ferguson, *History of the Improvement of the Lower Mississippi River for Flood Control and Navigation, 1932–1939*, iii.

26. "Report made to the Legislature on the State of Louisiana," *Mississippi Statesman and Natchez Gazette*, May 31, 1827.

27. Flood Control in the Mississippi Valley, 17; Humphreys and Abbott, Report upon the Physics and Hydraulics of the Mississippi River, 150–151, 155.

28. Camillo, *Divine Providence*, 145–48; Ferguson, *History of the Improvement of the Lower Mississippi River for Flood Control and Navigation, 1932–1939*, iii, 7; Final Report on Removing Wreck of Battleship "Maine" from Harbor of Habana, Cuba, 6; Reuss, *Designing the Bayous*, 137–38.

29. Hearings before the Committee on Flood Control, House of Representatives (1933), 27.

30. Ferguson, *History of the Improvement of the Lower Mississippi River for Flood Control and Navigation, 1932–1939*, iii, 7; Hearings before the Committee on Flood Control, House of Representatives (1933), 32–33; Morgan, *Dams and Other Disasters*, 228, 230, 232, 240, 246, 247–48; Reuss, *Designing the Bayous*, 148–50.

31. Ferguson, *History of the Improvement of the Lower Mississippi River for Flood Control and Navigation, 1932–1939*, 9, 177–80.

32. Camillo, *Divine Providence*, 149–150; Ferguson, *History of the Improvement of the Lower Mississippi River for Flood Control and Navigation, 1932–1939*, 15–16, 24, 163, 166; Hearings before the Committee on Commerce, United States Senate (1938), 49–50.

33. Camillo, *Divine Providence*, 159, 160; Ferguson, *History of the Improvement of the Lower Mississippi River for Flood Control and Navigation, 1932–1939*, 33, 45, 163; Hearings before the Committee on Flood Control, House of Representatives (1933), 31.

34. Hearings before the Committee on Commerce, United States Senate (1938), 49–50; Reonas, "Delta Planters and the Eudora Floodway," 173, 185.

35. Hearings before the Committee on Commerce, United States Senate (1938), 76–77, 80.

36. Barry, *Rising Tide*, 425; Biedenharn et al., "Recent morphological evolution of the Lower Mississippi River," 228, 231, 233, 243, 244, 246; Clauson, *Measuring Trends in Riverbed Gradation*, 4, 9, 41, 47.

37. Hearings before the Committee on Flood Control, House of Representatives (1933) 6, 19–20, 21, 57; Hebert, "Flood Control Capabilities of the Atchafalaya Basin Floodway," 2, 31, 43; Reuss, *Designing the Bayous*, 3, 5.

38. Hearings before the Committee on Flood Control, House of Representatives (1933) 9–10, 21, 23, 32–33, 74, 80, 79–80, 90; Hearings before the Committee on Flood Control, House of Representatives (1934), 58, 94–95; Hearings before the Subcommittee on Flood Control, United States Senate (1936), 33; Hebert, "Flood Control Capabilities of the Atchafalaya Basin Floodway," Figure 10, 31, 47–49, 57–67; Reuss, "The Army Corps of Engineers and Flood-Control Politics," 135.

39. Reuss, *Designing the Bayous*, 151–52.

40. Hebert, "Flood Control Capabilities of the Atchafalaya Basin Floodway," 2, 31.

41. Hearings before the Committee on Flood Control, House of Representatives (1933), 23; Hearings before the Committee on Flood Control, House of Representatives (1934), 78, 93–94; Hearings before the Subcommittee on Flood Control, United States Senate (1936), 179.

42. Hearings before the Committee on Flood Control, House of Representatives (1933), 70–73, 77; Hearings before the Committee on Flood Control, House of Representatives (1934), 53–54, 58, 74; Hearings before the Committee on Flood Control, US Senate (1936), 29, 33.

43. Humphreys and Abbott, Report upon the Physics and Hydraulics of the Mississippi River, 100, 153; "Object of Note," *Times-Picayune* (New Orleans), August 12, 1866; "The Inundation," *Times-Picayune* (New Orleans), April 26, 1874; "The Morganza Levee Surely Gone," *Times-Picayune* (New Orleans), April 21, 1874; "No Time To Be Discouraged," *Times-Picayune* (New Orleans), April 23, 1890; "The Inundation," *Times-Picayune* (New Orleans), April 26, 1874; "The Levees," *Times-Picayune* (New Orleans), December 12, 1874; Annual Report of the Mississippi River Commission for the Fiscal Year ending June 30, 1894; "Old River and the Mud Hole," *Times-Picayune* (New Orleans), September 18, 1877.

44. "A Fight with the River," *Times-Picayune* (New Orleans), May 3, 1890.

45. Hearings before the Subcommittee on Flood Control, United States Senate (1936), 265; Hearings before the Committee on Commerce, United States Senate (1938), 8, 80.

46. Hearings before the Committee on Flood Control, House of Representatives (1934), 73–74.

47. Annual Report of the Chief of Engineers, Part 1, vols. 1 and 2, 1941, 2178; Comprehensive Report on Reservoirs in Mississippi River Basin, 4–5; Flood Control Act of 1938, Section 2; O'Neill, *Rivers by Design*, 161–62; Reonas, "Delta Planters and the Eudora Floodway," 167, 177–78; Mary Linn Wernet, "The United States Senator Overton Collection and the History It Holds Relating to the Control of Floods in the Alluvial Valley of the Mississippi, 1936–1948," *Journal of the Louisiana Historical Association* 46, no. 4 (Autumn 2005): 462.

48. *Room for the River: Summary Report of the 2011 Mississippi River Flood and Successful Operation of the Mississippi River & Tributaries System* (n.p.: US Army Corps of Engineers and Mississippi River Commission, 2012), 2.

49. For a detailed account of the floodways saga, see Reuss, *Designing the Bayous*, 173–203.

50. Annual Report of the Chief of Engineers, 1940, Part 1, vols. 1 and 2, 2237; Hearings before the Committee on Commerce, United States Senate (1938), 36, 104, 148.

51. Hebert, "Flood Control Capabilities of the Atchafalaya Basin Floodway," Figure 8, 25, 43.

52. Annual Report of the Chief of Engineers, 1945, Part 1, vol. 2, 2456; Hearings before the Committee on Flood Control, House of Representatives (1945), 18–21; High Flows and Flood History on the Lower Mississippi River; Reuss, *Designing the Bayous*, 198–201.

53. Latimer and Schweizer, *Atchafalaya River Study*, vol. 3, Table 36. In the mid-1960s, after the Old River control structures were in place, low water on the Mississippi relative to the level of the Red River caused a near reversal of the current's direction through the low-sill gates. Perry Gustin, personal communication, January 19, 2015.

Chapter Five

1. Latimer and Schweizer, *Atchafalaya River Study*, vol. 3, Tables 13 and 14. A more recent estimate of daily sediment transport volume in the Lower Mississippi River averages about four hundred thousand tons. Thorne et al., *Current and Historical Sediment Loads in the Lower Mississippi River*, iii.

2. Robert Ettema and Cornelia F. Mutel, *Hans Albert Einstein: His Life as a Pioneering Engineer* (Reston, VA: American Society of Civil Engineers, 2014), 32–33, 55, 62, 94, 97.

3. Hans Albert Einstein, "Appendix II. Bed-Load Measurements in West Goose Creek" in "Bed-Load Transportation in Mountain Creek," Soil Conservation Service Report SCS-TP-55, US Department of Agriculture (Washington, DC: US Department of Agriculture, 1944), 46–50.

4. Harold N. Fisk, *Geological Investigation of the Alluvial Valley of the Lower Mississippi River*, Mississippi River Commission Publication No. 52, War Department, Corps of Engineers, US. Army (Vicksburg: Mississippi River Commission, 1944); Latimer and Schweizer, *Atchafalaya River Study*, vol. 1, Appendix D, 21–23.

5. Annual Report of the Chief of Engineers, 1949, Part 1, vol. 2, 2679; Latimer and Schweizer, *Atchafalaya River Study*, vol. 1, Appendix B, 28–29 and Appendix D, 17–19, 24–26, 37–38; C. P. Lindner, Memorandum Report on Protection of Red River Backwater by a Control Structure in Old River and Operation of Morganza Floodway and Bonnet Carré Spillway, Office of the President, Mississippi River Commission, Research and Technical Records Section, Mississippi River Commission Library, (Vicksburg, MS: Mississippi River Commission, August 13, 1945), 1, 4, Appendix 1–1; Reuss, *Designing the Bayous*, 214–15.

6. Lindner, Memorandum Report on Protection of Red River Backwater, Appendix 2–1.

7. High Flows and Flood History on the Lower Mississippi River; Lindner, Memorandum Report on Protection of Red River Backwater, Appendix 1–1, Appendix 2–2, 4; Latimer and Schweizer, *Atchafalaya River Study*, Vol. 1, Appendix D, 38.

8. Lindner, Memorandum Report on Protection of Red River Backwater, Appendix 2–2, 2, 4.

9. Annual Report of the Chief of Engineers, 1949, Part 1, vol. 2, 2682–83; Annual Report of the Chief of Engineers, 1950, Part 1, vol. 2, 2771; Annual Report of the Chief of Engineers, 1953, Part 1, vol. 2, 2034; Camillo, *Divine Providence*, 253.

10. Annual Report of the Chief of Engineers, 1949, Part 1, vol. 2, 3007; Latimer and Schweizer, *Atchafalaya River Study*, vol. 1, Appendix B, 7, Appendix D, 38; Morgan City 2011: Documenting the Floods for the Morgan City Archives, Tulane University School of Architecture, Service Learning, Fall 2011, 2, http://www.bk.psu

.edu/Documents/Academics/Keady-Molanphy_RWC.pdf (accessed March 23, 2015).

11. Latimer and Schweizer, *Atchafalaya River Study*, vol. 1, 43–44.

12. Ibid., vol. 1, 4–8.

13. Ibid., vol. 1, Appendix B, 5, 18, Appendix C, 3.

14. Not all of the Corps' engineers believed that the Mississippi was changing course. Brigadier General Hans Kramer, assistant Lower Mississippi Valley engineer, maintained that the Corps' dredging was responsible for the Atchafalaya's increased capacity and did not see a need for a control structure at Old River. Latimer and Schweizer, *Atchafalaya River Study*, vol. 1, Appendix D, 27–37; Reuss, *Designing the Bayous*, 215.

15. Latimer and Schweizer, *Atchafalaya River Study*, vol. 1, 21, 27, 39–40, Appendix B, 16, Appendix C, 6.

16. Annual Report of the Chief of Engineers, 1950, Part 1, vol. 2, 2770.

17. Latimer and Schweizer, *Atchafalaya River Study*, vol. 1, 15, 47; "Conferences to Consider Means of Preventing the Atchafalaya from Capturing the Mississippi," May 19, 1953 (no author), Folder 326, Box 51, Samuel D. Sturgis Jr. Papers, Sturgis Chief of Engineers 1953–1956, Office of History, USACE HQ, Alexandria, VA, 1.

18. Ettema and Mutel, *Hans Albert Einstein*, 193–96.

19. Hans Albert Einstein, *The Bed-Load Function for Sediment Transportation in Open Channel Flows*, Technical Bulletin No. 1026, September 1950 (Washington, DC: USDA Soil Conservation Service, 1950),

2–4, 14, 29–31, 67; Hsei–Wen Shen, "Appendix C. Hans Albert Einstein's Contributions to Hydraulics," in *Hans Albert Einstein: Reminiscences of His Life and Our Life Together*, by Elizabeth R. Einstein (Iowa City: Iowa Institute of Hydraulic Research, University of Iowa, 1991), 107.

20. Consultants Conference, Atchafalaya River Study, December 11–13, 1951, Minutes, Research and Technical Records, Library of the Mississippi River Commission, Vicksburg, MS, provided by IIHR-Hydroscience & Engineering, University of Iowa College of Engineering; Consultants Conference, Atchafalaya Structure Study, February 9, 1952, Research and Technical Records, Library of the Mississippi River Commission, Vicksburg, MS, provided by IIHR-Hydroscience & Engineering, University of Iowa College of Engineering, 1; Meeting to Consider Atchafalaya River Structures, Research and Technical Records, Library of the Mississippi River Commission, Vicksburg, MS, provided by IIHR-Hydroscience & Engineering, University of Iowa College of Engineering, August 25, 1952; Minutes of Meeting with Consultants on the Study of Means of Preventing the Atchafalaya River from Capturing the Mississippi River with Special Reference to Structures, September 29–30, 1952, Mississippi River Commission, Research and Technical Records, Library of the Mississippi River Commission, Vicksburg, MS, provided

by IIHR-Hydroscience & Engineering, University of Iowa College of Engineering; "Conferences to Consider Means of Preventing the Atchafalaya from Capturing the Mississippi," 4–5.

21. Consultants Conference, Atchafalaya Structure Study, February 9, 1952, 17; Meeting to Consider Atchafalaya River Structures, August 25, 1952, 6, 10–11, 42.

22. Consultants Conference, Atchafalaya Structure Study, February 9, 1952, 75.

23. Ettema and Mutel, *Hans Albert Einstein*, 80–82, 197.

24. Consultants Conference, Atchafalaya River Study, December 11–13, 1951, Minutes, Mississippi River Commission, 2–3; Ibid., February 9, 1952, 2, 10; Ettema and Mutel, *Hans Albert Einstein*, 188, 195–99.

25. Consultants Conference, Atchafalaya Structure Study, February 9, 1952, 4, 6, 9, 22; Minutes of Meeting with Consultants, September 29–30, 1952, 9.

26. Ewens, Red River Landing and Head of the Atchafalaya, 2638; C. R. Demas et al., "Sediment Transport in the Lower Mississippi River, 1983–1985," 4–86, in *Proceedings of the Fifth Federal Interagency Sedimentation Conference, 1991*, ed. Shou-Shan Fan and Yung-Huang Kuo, vol. 1, Interagency Advisory Committee on Water Data, Subcommittee on Sedimentation, Las Vegas, NV, March 18–21, 1991, 4–86.

27. The Clinton model was not fully completed until 1966 and Corps engineers used the expansive complex through the 1970s. Although the Corps' Waterways Experiment Station at Vicksburg began working with computer simulations in the mid-1950s, physical modeling continues alongside computational modeling, which examines water movement along one or more dimensions. "Atchafalaya Basin Floodway System," Louisiana: Feasibility Study, Technical Appendix C.1.7., US Army Corps of Engineers, Mississippi River Commission, New Orleans District, January 1982, https://archive.org/stream/AtchafalayaBasinFloodwaySystemLouisianaFeastilityStudyTechAppendices_212/AtchafalayaBasinFloodwaySystemLouisianaVolume2_djvu.txt (accessed September 16, 2014) ; Lower Mississippi River Sediment Study, Catalyst-Old River Hydroelectric Limited Partnership d/b/a Louisiana Hydroelectric Limited Partnership Vidalia, Louisiana in association with US Army Corps of Engineers New Orleans and Vicksburg Districts, US Army Corps of Engineers Waterways Experiment Station Coastal and Hydraulics Laboratory, Vicksburg, Mississippi, Colorado State University Engineering Research Center, Ft. Collins, Colorado, University of Iowa, Iowa Institute for Hydraulic Research, Iowa City, Iowa, Mobile Boundary Hydraulics, Clinton, Mississippi, May 28, 1999, available at the USACE ERDC Library, Vicksburg, Mississippi, vol. 1, 10; Ivan H. Nguyen, personal communication, July 15, 2015; Rollins interview, 145–46. Today, the Corps' concrete river-scape at Clinton lies nearly forgotten beneath an encroaching forest on the periphery of Buddy Butts Park, fenced-off and hidden from the driving range and soccer fields. A rusty water tower stands to mark the once-state-of-the-art facility. Kristi Dykema Cheramie, "The Scale of Nature:

Modeling the Mississippi River," Places Journal, March 2011, https://placesjournal. org/article/the-scale-of-nature-modeling-the-mississippi-river (accessed January 17, 2015); Consultants Conference, Atchafalaya Structure Study, February 9, 1952, 11; Ettema and Mutel, *Hans Albert Einstein*, 63, 81, 209; Meeting to Consider Atchafalaya River Structures, August 25, 1952, 16; Minutes of Meeting with Consultants, September 29–30, 1952, 15, 25.

28. Norman R. Moore, "Structures Required," 1172, in "Old River Diversion Control—A Symposium," Transaction of the American Society of Civil Engineers, 1958, 1172–81; Reuss, *Designing the Bayous*, 230, 240; "Conferences to Consider Means of Preventing the Atchafalaya from Capturing the Mississippi," 1–3.

29. See the discussion about point bar formation in Chapter 1. Consultants Conference, Atchafalaya Structure Study, February 9, 1952, 17–18, 36–41, 48–51, 56–58; Latimer and Schweizer, *Atchafalaya River Study*, vol. 1, Appendix C, 16; Meeting to Consider Atchafalaya River Structures, August 25, 1952, 30, 33, 35; Minutes of Meeting with Consultants, September 29–30, 1952, 25, 52.

30. According to Norman's Chart of the Lower Mississippi River by A. Persac (1858), a "Dr. Nock" owned a plantation here. Apparently, the name of the plantation's river landing site eventually changed from Nock's Landing to Knox Landing.

31. E. A. Graves, "Hydraulic Requirements," in "Old River Diversion Control—A Symposium," Transaction of the American Society of Civil Engineers, 1958, 1147; Minutes of Meeting with Consultants, September 29–30, 1952, 26–28; "Conferences to Consider Means of Preventing the Atchafalaya from Capturing the Mississippi," 3–5.

32. Consultants Conference, Atchafalaya River Study, December 11–13, 1951, 8–10; Meeting to Consider Atchafalaya River Structures, August 25, 1952, 1; Minutes of Meeting with Consultants, September 29–30, 1952, 12, 18–24.

33. John R. Harris, "Alternate Water Sources for the Baton Rouge-New Orleans Industrial Corridor, Addendum A, 4, Figure 1, in "If the Old River Control Structure Fails?," by Raphael G. Kazmann and David B. Johnson, *Louisiana Water Resources Research Institute*, Bulletin 12, September 1980, Louisiana State University, Baton Rouge; Hebert, "Flood Control Capabilities of the Atchafalaya Basin Floodway," 42, 50; Raphael G. Kazmann and David B. Johnson, "If the Old River Control Structure Fails? The Physical and Economic Consequences," *Louisiana Water Resources Research Institute*, Bulletin 12, September 1980, Louisiana State University, Baton Rouge, 24; Meeting to Consider Atchafalaya River Structures, August 25, 1952, 9, 10, 17, 18; Minutes of Meeting with Consultants, September 29–30, 1952, 17; *Old River Control*, US Army Corps of Engineers, New Orleans District, January 2009, 12; "Conferences to Consider Means of Preventing the Atchafalaya from Capturing the Mississippi," 2–4.

34. Consultants Conference, Atchafalaya Structure Study, February 9, 1952, 52; Meeting to Consider Atchafalaya River Structures, August 25, 1952, 6–7; Minutes of Meeting with Consultants, September 29–30, 1952, 2, 17; *Old River Control*, US Army Corps of Engineers, New Orleans District, January 2009, 9.

35. John R. Hardin, "Mississippi-Atchafalaya Diversion Problem," *Military Engineer* 46, no. 310 (March–April 1954): 92.

36. For more in-depth coverage of the maneuvering necessary to obtain relatively quick congressional approval for the Old River control project, see Reuss, *Designing the Bayous*, 234–38. Hardin, "Mississippi-Atchafalaya Diversion Problem," 92; Latimer and Schweizer, *Atchafalaya River Study*, vol. 1, 43–44; Minutes of First Meeting of Board of Consulting Engineers on Old River Control, April 27, 1954, Folder 326, Box 51, Samuel D. Sturgis Jr. Papers, Sturgis Chief of Engineers 1953–1956, Office of History, USACE HQ, Alexandria, VA; Reuss, *Designing the Bayous*, 230–31, 234–35, 238.

37. Flood Control Act of 1954, Public Law No. 780, 83rd Cong., Chapter 1264, 2d sess., H.R. 9859, September 3, 1954, http://planning.usace.army.mil/toolbox/library/PL/RHA1954.pdf (accessed September 16, 2014); Memorandum, Gail A. Hathaway, Office of the Chief of Engineers, to Major General Samuel D. Sturgis, Chief of Engineers, April 29, 1954, Folder 326, Box 51, Samuel D. Sturgis Jr. Papers, Sturgis Chief of Engineers 1953–1956, Office of History, USACE HQ, Alexandria, VA.; Minutes of First Meeting of Board of Consulting Engineers on Old River Control, April 27, 1954; Reuss, *Designing the Bayous*, 238, 240–41.

38. Graves, "Hydraulic Requirements," 1147, 1158; Old River Control Structure Sediment Diversion, 16; Reuss, *Designing the Bayous*, 229; Saucier, *Geomorphology and Quaternary Geologic History of the Lower Mississippi Valley*, Plate 10; Willard J. Turnbull and Woodland G. Shockley, "Foundation Design," in "Old River Diversion Control—A Symposium," Transaction of the American Society of Civil Engineers, 1958, Figures 1 and 2, 1163.

39. Annual Report of the Chief of Engineers, 1956, Part 1, vol. 2, 1870, 1879–1880; Annual Report of the Chief of Engineers, 1960, vol. 2, 1936; Annual Report of the Chief of Engineers, United States Army, 1959, vols. 1 and 2, 1923; Jennings, *Military Engineer*, 256.

40. The Corps added a second one-hundred-ton crane to the low sill after the 1973 flood. Perry Gustin, interviewed by James F. Barnett Jr., January 19, 2015, Morganza, Louisiana, 7, transcription in the author's research collection, 717 North Pearl Street, Natchez, MS.

41. Johnson, "A Change in the Course of the Lower Mississippi River, Addendum B," 4–5; *Old River Control*, US Army Corps of Engineers, New Orleans District, January 2009, 12; Turnbull and Shockley, "Foundation Design," 1169, 1176.

42. Robert T. Fairless, "The Old River Control Project," in *Proceedings, 27th Annual Conference and Symposium*, ed. Dhamo Dhamotharan and Harry C. McWreath (Bethesda, MD: American Water Resources Association), 266, Figure 3; Moore, "Structures Required," 1173, 1176, Figure 2; Reuss, *Designing the Bayous*, 244; *Old River Control*, 10; Turnbull and Shockley, "Foundation Design," 1170. The wing wall height is an estimate based upon the author's inspection of the low sill's remaining wing wall and comparisons of photographs with scale drawings of the structure, e.g., Johnson, "A Change in the Course of the Lower Mississippi River, Addendum B," 4.

43. Annual Report of the Chief of Engineers, 1957, vol. 2, 1844; Annual Report of the Chief of Engineers, 1960, vol. 2, 1936; Annual Report of the Chief of Engineers, 1959, Vols. 1 and 2, 1923; Gustin interview, 9; Hardin, "Mississippi-Atchafalaya Diversion Problem," 92; Johnson, "A Change in the Course of the Lower Mississippi River, Addendum B," 4–5; Moore, "Structures Required," Figure 6, 1179–80; Old River Control Structure Sediment Diversion, Hydraulic Model Investigation, Technical Memorandum No. 2–388, conducted for The President, Mississippi River Commission, Vicksburg, Mississippi, by Waterways Experiment Station, June 1954, 16; Reuss, *Designing the Bayous*, 230.

44. Andrew P. Rollins Jr. interviewed by Lynn M. Alperin, September 14–15, 1987, Dallas, Texas, transcript in Research Collections of the Office of History, Headquarters, US Army Corps of Engineers, Alexandria, VA, 279–80; Turnbull and Shockley, "Foundation Design," Figure 5. By comparison, the Port Allen lock just west of Baton Rouge, built around the same time as the Old River lock, is 84 feet wide. The Plaquemine lock, about 20 miles downriver from Port Allen, is only 55 feet wide. The two Intracoastal Waterway locks near New Orleans, at Harvey and Algiers, are both 75 feet wide. *Flood Control and Navigation Maps of the Mississippi River, 1960*, Sheet 1. On the upper Mississippi, New Lock 19 at Keokuk, Iowa, completed in 1957, is 110 feet wide. Fremling, *Immortal River*, 223. According to Morris Oubre, lockmaster at Old River lock, the 75-foot width causes "occasional, but seldom" problems with barge traffic. Oubre noted that standard barge widths vary from 35 to 40 feet. Personal communication, May, 14, 2015.

45. Annual Report of the Chief of Engineers, 1959, vols. 1 and 2, 1923; Annual Report of the Chief of Engineers, 1963, Part 1, vol. 2, 1799, 1809; Raymond Harrison, Corps of Engineers, Old River lock, personal communication, April 17, 2015; Edward B. Jennings, "The Life and Death of Old River," *Military Engineer* 6 (July–August 1964): 256 *Old River Control*, US Army Corps of Engineers, New Orleans District, January 2009, 9.

46. Hans Albert Einstein, Old River Control: Closure of Old River, Water Resources Archives, University of California Riverside and California State

University, San Bernadino Libraries, November 1953. 1, 6, 8; Jennings, "Life and Death of Old River," 257.

47. Jennings, "Life and Death of Old River," 256–57.

Chapter Six

1. High Flows and Flood History on the Lower Mississippi River; "The Watergate Story," *Washington Post*, http://www.washingtonpost.com/wp-srv/politics/special/watergate/timeline.html (accessed May 22, 2015).

2. Briefing-Mississippi River, Hearing Before the Committee on Public Works, House of Representatives, 93rd Cong., 1st sess., April 11, 1973 (Washington, DC: Government Printing Office, 1973), 2.

3. Moore, "Structures Required," 1180.

4. Jennings, "Life and Death of Old River," 256–57.

5. Annual Report of the Chief of Engineers, 1964, 47.

6. Minutes of Meeting with Consultants, September 29–30, 1952, 35.

7. Moore, "Structures Required," Figure 3.

8. Harris, "Alternate Water Sources for the Baton Rouge-New Orleans Industrial Corridor," 5; Hebert, "Flood Control Capabilities of the Atchafalaya Basin Floodway," Illustration 8, 50; Johnson, "A Change in the Course of the Lower Mississippi River," Figure 1.5.

9. Hebert, "Flood Control Capabilities of the Atchafalaya Basin Floodway," 42, 50; Charles R. Nickles and Thomas J. Pokrefke Jr., "Barge Barrier Study," Technical Report HL-84-4, Old River Diversion, Mississippi River, Report 3, Hydraulic Model Investigation," Hydraulic Laboratory, US Army Engineer Waterways Experiment Station, Vicksburg, MS, 1984, 5–6.

10. Nickles and Pokrefke, "Barge Barrier Study," 6–20; Annual Report of the Chief of Engineers on Civil Works Activities, vol. 1, 1972, 42–8; Annual Report of the Chief of Engineers on Civil Works Activities, vol. 2, 1970, 1224.

11. "Atchafalaya Basin Floodway System," A.7.7.

12. Hebert, "Flood Control Capabilities of the Atchafalaya Basin Floodway," 38–39.

13. Ibid., 3, 69, 71; Latimer and Schweizer, *Atchafalaya River Study*, vol. 1, 37.

14. Annual Report of the Chief of Engineers, 1973, 7–8; Briefing-Mississippi River, April 11, 1973, 4; Floods and Flood Control on the Mississippi, 1973, 1–2, US Army Corps of Engineers,
http://babel.hathitrust.org/cgi/pt?id=uiug.30112008443852;view=1up;seq=1.

15. Briefing-Mississippi River, April 11, 1973, 4; Floods and Flood Control on the Mississippi, 1973, 13–16; Charles C. Noble, interviewed by Martin Reuss,

September 22–23, 1981, transcript in Research Collections of the Office of History, Headquarters, US Army Corps of Engineers, Alexandria, VA, 124–27.

16. Floods and Flood Control on the Mississippi, 1973, 2, 21–23.

17. Ibid., 17–18.

18. Ibid., 26–27, 29, 31–32.

19. "Rains Bring Threat of More Floods," *Baton Rouge Sunday Advocate*, April 1, 1973.

20. "Mississippi River Crest Conditions Revised by Officials," *Baton Rouge Morning Advocate*, April 2, 1973.

21. Gustin interview, 2; "Floodways Explained," *Concordia Sentinel* (LA), March 28, 1973; "Rains Bring Threat of More Floods," *Baton Rouge Sunday Advocate*, April 1, 1973; "No Opening Planned Yet for Morganza Floodway," *Baton Rouge Morning Advocate*, April 3, 1973; "Flood Control A Profession," *Concordia Sentinel* (LA), April 11, 1973; "Solons Clash on Closing of Structure," *Concordia Sentinel* (LA), April 11, 1973; "Corps Adds Height to Floodwall to Protect Morgan City," *Times-Picayune* (New Orleans), April 12, 1973.

22. Floods and Flood Control on the Mississippi, 32; Lindner, Memorandum Report on Protection of Red River Backwater, Appendix 2–2, 2, 4.

23. Floods and Flood Control on the Mississippi, 35–37; "Spillway Opened Above New Orleans," *Baton Rouge Morning Advocate*, April 9, 1973; River Forecast, Lower Ohio/Mississippi River, Lower Mississippi River Forecast Center, National Weather Service, Slidell, LA.

24. Floods and Flood Control on the Mississippi, 1973, 36–37; "River Water to Take Toll In Seafood," *Baton Rouge Morning Advocate*, April 9, 1973. The Corps of Engineers, environmentalists, sport fishing interests, and the seafood industry continue to assess the Bonnet Carré spillway's impact on Lake Pontchartrain, Lake Borgne, and Mississippi Sound. During the flooding of December 2015-January 2016, a new round of environmental studies accompanied the spillway's opening. Although the Mississippi River water brings nutrients into the lakes and sound, the decomposition of introduced phytoplankton is harmful to fish populations. Some scientists also fear the introduction of the Asian carp into Lake Pontchartrain. Despite these concerns, coastal researchers note that the conditions caused by the use of the Bonnet Carré spillway can stabilize in eight to nine months. "Economic Impacts of the opening of Bonnet Carré Spillway to Mississippi Oyster Fisheries," Mississippi State University, Coastal Research and Extension Center, http://coastal.msstate.edu/oyster-bonnet-Carré (accessed March 1, 2016); "How would Bonnet Carré Spillway opening impact Pontchartrain fishing?," *Louisiana Sportsman*, January 5, 2016; "What the Bonnet Carré Spillway opening means for Lake Pontchartrain," *NOLA.com/The Times-Picayune*, January 8, 2016.

25. Briefing-Mississippi River, April 11, 1973, 11, 13, 19, 39; "Corps Prepares Morganza Area," *Times-Picayune* (New Orleans), April 11, 1973.

26. "Corps Prepares Morganza Area," *Times-Picayune* (New Orleans), April 11, 1973; Harris, "Alternate Water Sources for the Baton Rouge-New Orleans Industrial Corridor," 5–6; "Official Assesses High Water Concern," *Baton Rouge Morning Advocate*, April 14, 1973; Reuss, *Designing the Bayous*, 243.

27. Gustin interview, 3, personal communication, June 17, 2015.

28. Russell A. Beauvais, personal communication, June 22, 2015.

29. Gustin interview, 5, 12; Harris, "Alternate Water Sources for the Baton Rouge-New Orleans Industrial Corridor," 5; Noble interview, 130. Confusingly, different sources give different dates for the low-sill's wing wall collapse. A *Times-Picayune* article published on Monday, April 16, quotes General Noble as saying that the Corps became aware of a structural problem with the low sill's south fore bay wing wall on Thursday, April 12. "Morganza To Open Tomorrow—Corps," *The Times-Picayune* (New Orleans), April 16, 1973. The Corps' Mississippi River and Tributaries Post-Flood Report for 1973 states that the wing wall was undermined and collapsed on Thursday, April 12. *Mississippi River and Tributaries Post-Flood Report, 1973*, 8, Department of the Army Corps of Engineers, Lower Mississippi Valley Division, Vicksburg, MS. However, the Mississippi River Commission's 1973 flood report in that year's Annual Report of the Chief of Engineers gives the wing wall collapse date as Friday evening, April 13. Annual Report of the Chief of Engineers on Civil Works Activities, vol. 1, 1973, 42–47. The Monday, April 16 edition of the *Baton Rouge Advocate* quotes an unnamed Corps of Engineers spokesperson as saying that the wing wall fell on Sunday, April 15. "Corps Plans to Open Spillway at Morganza," *Baton Rouge Morning Advocate*, April 16, 1973. I have based my narrative of events on remarks by General Noble on Monday, April 16, at a press conference in Monroe, Louisiana, where he stated twice that the wing wall failed "late Saturday night" (April 14) and that the Corps opened the overbank structure on Sunday, April 15. "River Threatens Structure," *Concordia Sentinel* (LA), April 18, 1973. General Noble's statements along with several other sources indicate that April 14 is the correct date for the wing wall collapse, and that the structure's failure was unforeseen before that night. "Corps Plans to Open Spillway at Morganza," *Baton Rouge Morning Advocate*, April 16, 1973; Noble interview, 125–130; Fairless, "Old River Control Project," 266; Gustin interview, 6–8; Johnson, "A Change in the Course of the Lower Mississippi River," 5.

30. Gustin interview, 3–6, 12.

31. Ibid., 6.

32. Ibid., 2–7; Perry Gustin, personal communication, June 17, 2015.

33. Gustin interview, 6–8.

34. Ibid., 8–9; Gustin, personal communication, June 17, 2015; Harris, "Alternate Water Sources for the Baton Rouge-New Orleans Industrial Corridor," 5–6; John McPhee, "Atchafalaya," in *The Control of Nature* (New York: Farrar, Straus and Giroux, 1989), 29–30.

35. Gustin interview, 9–10; Reuss, *Designing the Bayous*, 244–245; Noble interview, 128–29.

36. After the 1973 flood, the Corps equipped the low sill with a second gantry crane.

37. Gustin interview, 10.

38. Gustin, personal communication, February 17, 2015; Gustin interview, 10–11.

39. Gustin interview, 11; Johnson, "A Change in the Course of the Lower Mississippi River," 4; Moore, "Structures Required," Figure 3; *Old River Control*, 12.

40. Gustin interview, 11–13; Noble interview, 129.

41. "Morganza Levees Will Be Bolstered," *Times-Picayune* (New Orleans), April 15, 1973; "River Menace Diminishes," *Times-Picayune* (New Orleans), April 15, 1973.

42. Gustin interview, 17; "River Threatens Structure," *Concordia Sentinel* (LA), April 18, 1973; "Morganza To Open Tomorrow—Corps," *Times-Picayune* (New Orleans), April 16, 1973; Noble interview, 128–29, 131–32.

43. "Corps Plans to Open Spillway at Morganza," *Baton Rouge Morning Advocate*, April 16, 1973; "Morganza To Open Tomorrow—Corps," *Times-Picayune* (New Orleans), April 16, 1973; Homer Willis, interviewed by Bruce Kalk, March 15, 1991, Bethesda, MD, Transcript in Research Collections of the Office of History, Headquarters, US Army Corps of Engineers, Alexandria, VA., 126.

44. Noble interview, 131.

45. "Corps Plans to Open Spillway at Morganza," *Baton Rouge Morning Advocate*, April 16, 1973; Floods and Flood Control on the Mississippi, 1973, 37; Noble interview, 132.

46. Floods and Flood Control on the Mississippi, 1973, 40.

47. Russell A. Beauvais, personal communication June 23, 2015; "Agreement on Spillway to Limit Water Flow," *Baton Rouge Morning Advocate*, April 17, 1973; "Morganza Use Is Reduced to One-Third Its Capacity," *Times-Picayune* (New Orleans), April 17, 1973; Willis interview, 127; Annual Report of the Chief of Engineers, 1973, 15.

48. "Corp [*sic*] Stands Vigil As Mississippi Rises," *Concordia Sentinel* (LA), April 18, 1973; "River Threatens Structure," *Concordia Sentinel* (LA), April 18, 1973.

49. "River Threatens Structure," *Concordia Sentinel* (LA), April 18, 1973.

50. "Corps Raising Backwater Levees," *Concordia Sentinel* (LA), May 16, 1973; "Engineers Confident In Backwater Levee," *Concordia Sentinel* (LA), May 2, 1973; Gustin interview, 24–25.

51. Fairless, "Old River Control Project," 266; "River Threatens Structure," *Concordia Sentinel* (LA), April 18, 1973.

52. "Little Man's Corner, Timber Cut in Flood Area," *Concordia Sentinel* (LA), May 23, 1973.

53. Noble interview, 126.

54. Ibid.

55. Floods and Flood Control on the Mississippi, 1973, 61.

56. High Flows and Flood History on the Lower Mississippi River.

57. Harris, "Alternate Water Sources for the Baton Rouge-New Orleans Industrial Corridor," 5.

58. Kazmann and Johnson, "If the Old River Control Structure Fails?," 20.

59. Camillo, *Divine Providence*, 228–31; Floods and Flood Control on the Mississippi, 1973, 39–40, 47; High Flows and Flood History on the Lower Mississippi River.

60. Fairless, "Old River Control Project," 266, Figure 4; Harris, "Alternate Water Sources for the Baton Rouge-New Orleans Industrial Corridor," 6, Figures 1 and 2; Hebert, "Flood Control Capabilities of the Atchafalaya Basin Floodway," Illustration 8.

61. Gustin interview, 20.

Chapter Seven

1. Kazmann and Johnson, "If the Old River Control Structure Fails," 8.

2. Kazmann and Harris were professors of civil engineering; Johnson was a professor of economics.

3. Kazmann and Johnson, "If the Old River Control Structure Fails," 1.

4. Noble interview, 128.

5. Kazmann and Johnson, "If the Old River Control Structure Fails," 4, 8, 40.

6. Ibid., 1.

7. Kazmann and Johnson, "If the Old River Control Structure Fails," 10–16.

8. Harris, "Alternate Water Sources for the Baton Rouge-New Orleans Industrial Corridor," 17; Kazmann and Johnson, "If the Old River Control Structure Fails," 5, 10–12.

9. Johnson, "A Change in the Course of the Lower Mississippi River, Addendum B," 8, 34; Kazmann and Johnson, "If the Old River Control Structure Fails," 1, 28, 41.

10. Gulf South Pipeline Company; Overview, http://www.gulfsouthpl.com/AboutUsGS.aspx (accessed February 5, 2013); Johnson, "A Change in the Course of the Lower Mississippi River, Addendum B,"

19; Kazmann and Johnson, "If the Old River Control Structure Fails," 29–32; Map: Texas Gas Transmission, LLC, Boardwalk Pipeline Partners, LP, http://www.txgt.com/Safety.aspx?id=310 (accessed February 5, 2013); National Pipeline Mapping system, Public Map Viewer, https://www.npms.phmsa.dot.gov/PublicViewer (accessed July 24, 2015); Nussbaum, *Louisiana Electric Generation–2007 Update*. Technology Assessment Division, Louisiana Department of Natural Resources, 1–2. South Louisiana Pipelines, Department of Natural Resources, State of Louisiana; US Energy Information Administration; Natural Gas; US Department of Energy.

11. David K. Schneider, "A Matter of Time-Eastern Pipeline Capacity," *We Are The Practitioners: We are Supply Chain Coaches*, March 26, 2012.

12. Johnson, "A Change in the Course of the Lower Mississippi River, Addendum B," 9; Kazmann and Johnson, "If the Old River Control Structure Fails," 26–27.

13. Johnson, "A Change in the Course of the Lower Mississippi River, Addendum B," 9; Kazmann and Johnson, "If the Old River Control Structure Fails," 26–27.

14. Campanella, *Bienville's Dilemma*, geographical insert: Population by Parish; Nussbaum, *Louisiana Electric Generation—2007 Update*, 1–2, Figure 2; Kazmann and Johnson, "If the Old River Control Structure Fails," 12.

15. Nussbaum, *Louisiana Electric Generation–2007 Update*, 1–2, Technology Assessment Division, Louisiana Department of Natural Resources (Baton Rouge: LDNR, 2007), 1–2, Figure 2.

16. Kazmann and Johnson, "If the Old River Control Structure Fails," 11, 14.

17. Campanella, *Bienville's Dilemma*, 227–28; Harris, "Alternate Water Sources for the Baton Rouge-New Orleans Industrial Corridor," 10; Kazmann and Johnson, "If the Old River Control Structure Fails," 14, 25.

18. *Thalweg* is a German word meaning "valley way."

19. Harris, "Alternate Water Sources for the Baton Rouge-New Orleans Industrial Corridor," Figure 3, 10–11; Kazmann and Johnson, "If the Old River Control Structure Fails," 21, 24.

20. Robert H. Mead, "Setting: Geology, Hydrology, Sediments, and Engineering of the Mississippi River: Water Discharge," US Geological Survey Circular 1133, Contaminants in the Mississippi River, Reston, VA, 1995, http://pubs.usgs.gov/circ/circ1133/geosetting.html (accessed July 31, 2015).

21. Kazmann and Johnson, "If the Old River Control Structure Fails," 14, 25.

22. Harris, "Alternate Water Sources for the Baton Rouge-New Orleans Industrial Corridor," 1; Kazmann and Johnson, "If the Old River Control Structure Fails," 25, 37; B. Pierre Sargent, *Water Use in Louisiana, 2010*, Water Resources Special Report No. 17 (2011; rev. ed., Baton Rouge: Louisiana Department of Transportation and Development, December 2012), 110, 111, 119, 125.

23. Harris, "Alternate Water Sources for the Baton Rouge-New Orleans Industrial Corridor," 12–14, 21, 43; Kazmann and Johnson, "If the Old River Control Structure Fails," 14, 15; Sargent, *Water Use in Louisiana, 2010*, 110. The 2010 percentages given refer only to the Baton Rouge-New Orleans corridor; total water usage by the Bayou Lafourche communities in 2010 amounted to less than 1 percent of the overall Mississippi water consumption below Baton Rouge. Ibid., 111.

24. Kazmann and Johnson, "If the Old River Control Structure Fails," Tables 1A and 1B.

25. Johnson, "A Change in the Course of the Lower Mississippi River, Addendum B," 8; Kazmann and Johnson, "If the Old River Control Structure Fails," 15, 29–30, 41.

26. Kazmann and Johnson, "If the Old River Control Structure Fails," 4, 35–36, 42.

27. Ibid., 21.

28. Campanella, *Bienville's Dilemma*, 227–28; D. C. Dial and D. M. Sumner, Geohydrology and Simulated Effects of Pumpage on the New Orleans Aquifer System at New Orleans, Louisiana, Water Resources Technical Report No. 46, Louisiana Department of Transportation and Development in cooperation with the US Geological Survey, 1989, 2, 51; Draft Programmatic Environmental Assessment, New Orleans Sewerage and Water Board Facilities and Carrollton Water Treatment Plant Hazard Mitigation Proposals, Orleans Parish, Louisiana, HMGP Multiple Projects, FEMA-1603-DR-LA, US Department of Homeland Security, New Orleans, Louisiana Recovery Office, March 2015, 4; Harris, "Alternate Water Sources for the Baton Rouge-New Orleans Industrial Corridor," 22, 27–28; Lawrence B. Prakken, Groundwater Resources in the New Orleans Area, 2008, Water Resources Technical Report No. 80, US Department of the Interior, US Geological Survey in cooperation with the Louisiana Department of Transportation and Development, Baton Rouge, LA 2009, Table 3, 33–34.

29. Colton, *An Unnatural Metropolis*, 64; Harris, "Alternate Water Sources for the Baton Rouge-New Orleans Industrial Corridor," 17–19, 22, 26, 28; "Tensions rise over salt water intrusion into BR aquifers," *Baton Rouge Advocate*, May 5, 2014.

30. Harris, "Alternate Water Sources for the Baton Rouge-New Orleans Industrial Corridor," 23, 28–30.

31. Colton, *An Unnatural Metropolis*, 127, 130–37.

32. Mark S. Davis, personal communication, July 30, 2015.

33. Harris, "Alternate Water Sources for the Baton Rouge-New Orleans Industrial Corridor," 23–24; Act No. 581, Regular Session, 2008 Louisiana State Legislature House Bill No. 376 by Representatives Dove and St. Germain, to enact R.S.

38:3097.3 (E), relative to ground water management; Act No. 955, Regular Session, 2010 Louisiana State Legislature House Bill No. 1486 (Substitute for House Bill No. 926 by Representative Little) by Representatives Morris et al. relative to the use of surface water, https://www.legis.la.gov/Legis/ViewDocument.aspx?d=722976 (accessed August 5, 2015); James M. Klebba, "Water Rights and Water Policy in Louisiana: Laissez Faire Riparianism, Market Based Approaches, or a New Managerialism?" *Louisiana Law Review* 5, no. 6 (July 1993): 1779.

34. Craig E. Colton, review of *Beyond Control* manuscript for University Press of Mississippi, February 2016, University Press of Mississippi, Jackson.

35. Kazmann and Johnson, "If the Old River Control Structure Fails," 5.

Chapter Eight

1. Noble interview, 127–128.

2. Fairless, "Old River Control Project," 266, 268; Noble interview, 129.

3. Fairless, "Old River Control Project," 266; Gustin interview, 19–20.

4. Atchafalaya Basin Floodway System, Louisiana: Feasibility Study, US. Army Corps of Engineers, Mississippi River Commission, New Orleans District, January 1982, Technical Appendix A-26, A-236; Fairless, "Old River Control Project," 267; Gustin interview, 8; John D. Higby Jr., *Possible Capture of the Mississippi River by the Atchafalaya River*, Colorado Water Resources Research Institute, Information Series No. 50, Colorado State University, AE 695V Special Study, August 1983, 43.

5. Atchafalaya Basin Floodway System, Louisiana: Feasibility Study, Technical Appendix A-235; Higby, *Possible Capture of the Mississippi River by the Atchafalaya River*, 43, 45.

6. NGVD (National Geodetic Vertical Datum) replaced Mean Sea Level in the 1970s as a more accurate base from which to measure vertical distances. In 1988, NAVD (North American Vertical Datum) replaced NGVD. Vertical Datums, National Geodetic Survey, National Oceanic and Atmospheric Administration, http://www.ngs.noaa.gov/datums/vertical (accessed July 7, 2015).

7. Fairless, "Old River Control Project," 268, Figure 5; Fletcher and Bhramayana, "Old River Control Auxiliary Structure, Hydraulic Model Investigation, 1, 5, Plates 6 and 27; Higby, *Possible Capture of the Mississippi River by the Atchafalaya River*, 46.

8. *Design of Spillway Tainter Gates*, 1–1, 2–1, 2–2, Table D-1; B. P. Fletcher and P. Bhramayana, "Old River Control Auxiliary Structure, Hydraulic Model Investigation, Technical Report HL-88–14, Hydraulics Laboratory, Department of the Army, Waterways Experiment Station, Corps of Engineers, Vicksburg, MS, Final Report, June 1988, 1, 5.

9. Nguyen et al., *Mississippi River and Old River Control Complex Sediment Investigation and Hydraulic Sediment Response Model Study*, Technical Report M53, US Army Corps of Engineers, St. Louis District, Hydrologic and Hydraulics Branch, Applied River Engineering Center, Final Report, March 2011, 9.

10. Fairless, "Old River Control Project," 261, 268; *Old River Control*, 12; Higby, *Possible Capture of the Mississippi River by the Atchafalaya River*, 45–46; High Flows and Flood History on the Lower Mississippi River; Kazmann and Johnson, "If the Old River Control Structure Fails," 34; Nickles and Pokrefke, "Barge Barrier Study," 20.

11. Reuss, *Designing the Bayous*, 247.

12. Fairless, "Old River Control Project," 261; "Hydro-electric plant, Engineers optimistic," *Concordia Sentinel* (LA), February 7, 1978; "Vidalia, Police Jury join forces to seek industry," *Concordia Sentinel* (LA), November 29, 1977; "Vidalia power plant nearer to reality," *Concordia Sentinel* (LA), February 23, 1978; "Hydro christening set," *Concordia Sentinel* (LA), March 29, 1989; "A new era in energy begins with Vidalia hydro plant," *Concordia Sentinel* (LA), October 14, 1985.

13. *Old River Control*, 15.

14. "1 million hydro rebate paid," *Concordia Sentinel* (LA), March 26, 1987; "Hydro christening set," *Concordia Sentinel* (LA), March 29, 1989.

15. The "vessel" was named for the Merrimac Corporation, owned by Thomas B. Pickens III, which was Catalyst's parent company. "Hydro christening set," *Concordia Sentinel* (LA), March 29, 1989.

16. *Old River Control*, 15.

17. "Hydro christening set," *Concordia Sentinel* (LA), March 29, 1989; *Old River Control*, 15.

18. Lower Mississippi River Sediment Study, vol. 1, 7, 8, 17; Nguyen et al., *Mississippi River and Old River Control Complex Sediment Investigation*, 9–10.

19. Joe Harvey, personal communication, October 6, 2011; Russell Beauvais, personal communication, August 17, 2015.

20. Nguyen et al., *Mississippi River and Old River Control Complex Sediment Investigation*, 8; Sargent, *Water Use in Louisiana, 2010*, 133.

21. "Startups explore alternative hydro power on the Mississippi." *Midwest Energy News* (St. Paul, MN), June 21, 2011.

22. Fremling, *Immortal River*, 17; "Free Flow Power seeks approval for hydropower plants on Allegheny, Monongahela, and Ohio Rivers." *Pittsburgh Post-Gazette*, April 1, 2014; "Free Flow Power hydrokinetic turbine deployed in Mississippi River." *Hydro Review* (Tulsa, OK), HydroWorld.com, July 13, 2011; "Startups explore alternative hydro power on the Mississippi." *Midwest Energy News* (St. Paul, MN), June 21, 2011.

23. Mississippi River and Tributaries Project, Mississippi River Mainline Levees, Enlargement and Seepage Control, US Army Corps of Engineers, Vicksburg District, Vicksburg, MS, Project Report, July 1998,

8–10, Appendix 4, Plate 4; Riparian Damages on the East Bank of the Mississippi River, Hearings before the United States Congress, March 22, 1912, 60.

24. Camillo, *Divine Providence*, 251–52.

25. High Flows and Flood History on the Lower Mississippi River.

26. Atchafalaya Basin Floodway System, Louisiana: Feasibility Study, Technical Appendix A-237; Camillo, *Divine Providence*, 161–62; Higby, *Possible Capture of the Mississippi River by the Atchafalaya River*, 46–52; Kazmann and Johnson, "If the Old River Control Structure Fails," 18–20; Lower Mississippi River Sediment Study, vol. 1, 17.

27. Atchafalaya Basin Floodway System, Louisiana: Feasibility Study, Technical Appendix A-23; Camillo, *Divine Providence*, 10, 265; Hebert, "Flood Control Capabilities of the Atchafalaya Basin Floodway," 38; Kazmann and Johnson, "If the Old River Control Structure Fails," 19–20, 40–41; Mississippi River and Tributaries Project, Mississippi River Mainline Levees, 8–10.

28. Atchafalaya Basin Floodway System, Louisiana: Feasibility Study, Technical Appendix A-50, 53; Bryan P. Piazza, *The Atchafalaya Basin: History and Ecology of an American Wetland* (n.p.: Nature Conservancy, 2014), 153.

29. Atchafalaya Basin Floodway System, Louisiana: Feasibility Study, Technical Appendix A-23, 25, 145; Camillo, *Divine Providence*, 236; Meeting to Consider Atchafalaya River Structures, August 25, 1952, 18.

30. Karen Dinicola, The "100-Year Flood," US Geological Survey Fact Sheet 229-96, last modified February 18, 2014; http://pubs.usgs.gov/fs/FS-229-96 (accessed October 31, 2014).

31. Gilbert F. White, *Human Adjustment to Floods: A Geographical Approach to the Flood Problem in the United States*, Research Paper No. 29, University of Chicago, Department of Geography Research Papers (Chicago: University of Chicago, 1945), 2.

32. Moore and Moore, *Army Corps of Engineers and the Evolution of Federal Flood Plain Management Policy*, 35–37, 45; White, *Human Adjustment to Floods*, 50.

33. Moore and Moore, *Army Corps of Engineers and the Evolution of Federal Flood Plain Management Policy*, 50–51, 53, 60.

34. Floods and Flood Control on the Mississippi, 1973, 57; Kazmann and Johnson, "If the Old River Control Structure Fails," 20; Moore and Moore, *The Army Corps of Engineers and the Evolution of Federal Flood Plain Management Policy*, 41–42, 71, 101–2.

35. All-Hazard Authorities of the Federal Emergency Management Agency; the National Flood Insurance Act of 1968, as amended, and the Flood Disaster

Protection Act of 1973, as amended 42 U.S.C. 4001 *et seq.*, Office of the General Counsel, August 1997, https://www.fema.gov/media-library/assets/documents/7277 (accessed March 12, 2016); Damon P. Coppola, *Introduction to International Disaster Management*, 3rd ed. (Burlington, MA: Butterworth-Heinemann/Elsevier, 2015), 259; Federal Emergency Management Agency, Flood Zones, National Flood Insurance Program Policy Index, http://www.fema.gov/floodplain-management/flood-zones (accessed October 31, 2014); H. C. Higgs, *Frequency Curves*, Techniques of Water-Resource Investigations of the United States Geological Survey, Chapter A2, Book 4, Hydrologic Analysis and Interpretation, Department of the Interior (Washington, DC: Government Printing Office, 1968, 1, 3, 8, 13, 14; High Flows and Flood History on the Lower Mississippi River; Moore and Moore, *Army Corps of Engineers and the Evolution of Federal Flood Plain Management Policy*, 74–75; Richard E. Sparks, "Rethinking, Then Rebuilding New Orleans," *Issues in Science and Technology* 22, no. 2 (Winter 2006): 7, 11, http://issues.org/22-2/sparks/ (accessed September 24, 2015).

36. Roy B. Van Arsdale, *Adventures Through Deep Time: The Central Mississippi River Valley and Its Earthquakes*, Geological Society of America, Special Papers 455 (Boulder, C): GSA, 2009), 83, 86–87.

37. Charles M. Elliott et al., "Response of Lower Mississippi River Low-Flow Stages," in 4-18–4-19, in *Proceedings of the Fifth Federal Interagency Sedimentation Conference, 1991*, vol. 1, ed. Shou-Shan Fan and Yung-Huang Kuo, Interagency Advisory Committee on Water Data, Subcommittee on Sedimentation, Las Vegas, NV, March 18–21, 1991, 4-16—4-19; Ellis L. Krinitzsky and Harold N. Fisk, Geological Investigation of Faulting in the Lower Mississippi Valley, Technical Memorandum No. 3-311, Waterways Experiment Station, Vicksburg, Mississippi, May 1950, 1, 45, Appendix 8-17, 34–36, 39; Saucier, *Geomorphology and Quaternary Geologic History of the Lower Mississippi Valley*, 302–4.

38. Saucier, *Geomorphology and Quaternary Geologic History of the Lower Mississippi Valley*, Figure 6, 52, 63, 65–66, 305; David T. Dockery III, *Windows into Mississippi's Geologic Past.*, Circular 6, Mississippi Department of Environmental Quality, Office of Geology (Jackson, MS: Department of Environmental Quality, 1997), 9, Figures 4 and 5; Higby, *Possible Capture of the Mississippi River by the Atchafalaya River*, 46, 53–55; Kurt D. Shinkle and Roy K. Dokka, Rates of Vertical Displacement at Benchmarks in the Lower Mississippi Valley and the Northern Gulf Coast, NOAA Technical Report NOS/NGS 50, US Department of Commerce, National Oceanic and Atmospheric Administration, National Ocean Service, July 2004, 1.

39. Ibid., 1, 3, Figure 5; Atchafalaya Basin Floodway System, Louisiana: Feasibility Study, Technical Appendix A-74–75; Campanella, *Bienville's Dilemma*, 54, 205, 324–29; Dockery, *Windows into Mississippi's Geologic Past*, Figure 5,

Appendix 1; Kazmann and Johnson, "If the Old River Control Structure Fails," 8; Krinitzsky and Fisk, Geological Investigation of Faulting in the Lower Mississippi Valley, 20, 36, Appendix 3–4, Figure A1; Saucier, *Geomorphology and Quaternary Geologic History of the Lower Mississippi Valley*, 51, 53, Table 2.

40. Campanella, *Bienville's Dilemma*, 324.

41. Shinkle and Dokka, Rates of Vertical Displacement at Benchmarks in the Lower Mississippi Valley and the Northern Gulf Coast, 6, 9; N. L. Bindoff et al., "Oceanic Climate Change and Sea Level," in *Climate Change 2007: The Physical Science Basis, Contribution of Working Group I to the Fourth Assessment Report of the Intergovernmental Panel on Climate Change*, ed. S. Solomon, D. Qin, M. Manning, Z. Chen et al. (Cambridge: Cambridge University Press, 2007), 409.

42. Guidance for Incorporating Climate Change Impacts to Inland Hydrology in Civil Works Studies, Designs, and Projects, US Army Corps of Engineers, *Engineering and Construction Bulletin No. 2014–10*, May 2014, https://www.wbdg.org/ccb/ARMYCOE/COEECB/ecb_2014_10.pdf (accessed August 25, 2015), 1–2.

43. Ibid., Appendix C-5.

44. Kenneth E. Kunkel et al., Climate of the Midwest, Regional Climate Trends and Scenarios for the US National Climate Assessment, Part 3, US National Oceanic and Atmospheric Administration Technical Report NESDIS (Washington, DC: US Department of Commerce, National Environmental Satellite, Data, and Information Service, June 2013) 142–43, Figures 31, 42, 44.

45. Kenneth E. Kunkel et al., Climate of the US Great Plains, Regional Climate Trends and Scenarios for the US National Climate Assessment, Part 4, US National Oceanic and Atmospheric Administration Technical Report NESDIS (Washington, DC: US Department of Commerce, National Environmental Satellite, Data, and Information Service, January 2013), 142–44, Figures 15, 16, 26.

46. Kenneth E. Kunkel et al., Climate of the Southeast, Regional Climate Trends and Scenarios for the US National Climate Assessment, Part 2, US National Oceanic and Atmospheric Administration Technical Report NESDIS 142-2, US Department of Commerce, National Environmental Satellite, Data, and Information Service, Washington DC, June 2013, Figures 28, 39, 48, 49.

Chapter Nine

1. Lower Mississippi River Sediment Study, vol. 1, 19–62; Nguyen et al., Mississippi River and Old River Control Complex Sediment Investigation, 1, 10–14, 38.

2. Nguyen et al., Mississippi River and Old River Control Complex Sediment Investigation, 1, 11, 16, 18, 24–38.

3. Davinroy et al., Hydraulic Sediment Response Modeling, Replication Accuracy to the River, Technical Paper M53, US Army Corps of Engineers, St. Louis and Memphis Districts, August 2011, 1–2, 16, http://mvs-wc.mvs.usace.army.mil/arec/Documents/Publications/M53_Hydraulic_Sediment_Response_Modeling_Replication_Accuracy.pdf (accessed July 11, 2015); HSR Modeling Theory, Applied River Engineering Center, St. Louis District, USACE, http://mvs-wc.mvs.usace.army.mil/arec/HSR_Modeling_Theory.html (accessed July 11, 2015); Nguyen et al., Mississippi River and Old River Control Complex Sediment Investigation, 11, 25.

4. Nguyen et al., Mississippi River and Old River Control Complex Sediment Investigation, 15, 18–20, 38–39.

5. Ibid., 18–19, 28–30, 32, Plate 34.

6. Minutes of Meeting with Consultants, September 29–30, 1952, 26–28.

7. Nguyen et al., Mississippi River and Old River Control Complex Sediment Investigation, 41.

8. Ibid., 32, 38–39.

9. Ibid., 40–41.

10. Camillo, *Divine Providence*, 251–52; Harris, "Alternate Water Sources for the Baton Rouge-New Orleans Industrial Corridor," 5; High Flows and Flood History on the Lower Mississippi River; Kazmann and Johnson, "If the Old River Control Structure Fails?," 20; Reuss, *Designing the Bayous*, 243.

11. Camillo, *Divine Providence*, 257, 261.

12. High Flows and Flood History on the Lower Mississippi River.

13. Dredging Activities FY2008-FY2015, New Orleans District, US Army Corps of Engineers, http://www.mvn.usace.army.mil/Missions/Navigation/DredgingInformation.aspx (accessed August 31, 2015); Joe Harvey, personal communication, August 31, 2015); Russell Beauvais, personal communications, August 31, 2015, September 4, 2015, and September 8, 2015.

14. Nguyen, personal communication, July 15, 2015.

15. Martin Hoerling et al., *An Interpretation of the Origins of the 2012 Central Great Plains Drought. Assessment Report*, NOAA Drought Task Force in partnership with the National Integrated Drought Information System, March, 20, 2013, 1, http://www.drought.gov/media/pgfiles/2012-Drought-Interpretation-final.web-041013_V4.0.pdf (accessed August 31, 2015).

16. "2012 drought impacting Mississippi River barges," *AccuWeather.com*, July 25, 2012, http://www.accuweather.com/en/weather-news/drought-impacting-mississippi-barges/67963 (accessed August 28, 2015).

17. "Corps begins building barrier to block saltwater in Mississippi River." *Fox 8, WVUE New Orleans*, August 30, 2012, http://www.fox8live.com/story/19300284/update-army-corps-begins-building-underwater-barrier-to-block-salt-water-in

-mississippi-river?clienttype=printable (accessed September 5, 2015); "Drought sends Mississippi into uncharted 'territory,'" *NBC News*, August 15, 2012; "Underwater dam in Mississippi River still protecting New Orleans, Jefferson water supplies from saltwater as drought continues." *Times-Picayune* (New Orleans), NOLA. com, December 17, 2012.

18. "After drought, reducing water flow could hurt Mississippi River transport," *New York Times*, November 26, 2012; "Drought sends Mississippi into uncharted 'territory,'" *NBC News*, August 15, 2012; "Gavins Point releases begin scheduled decrease as navigation season closes," Northwestern Division, US Army Corps of Engineers News Release December 7, 2012, http://www.nwd. usace.army.mil/Media/NewsReleases/tabid/1989/Article/475773/gavins-point-releases-begin-scheduled-decrease-as-navigation-season-closes.aspx (accessed September 3, 2015); "How the drought on the Mississippi River impacts you," *National Geographic News*, February 1, 2013, http://news.nationalgeographic.com/ news/2012/12/121207-nation-mississippi-river-drought-environment-economy (accessed August 28, 2015).

19. Camillo, *Divine Providence*, 3, 23.

20. Hoerling et al., *An Interpretation of the Origins of the 2012 Central Great Plains Drought. Assessment Report*, 1, 23–25, 29–30, 38.

21. Lower Mississippi River Sediment Study, vol. 1, 9, 17, vol. 6, Figure 4.

22. Ewens, Red River Landing and Head of the Atchafalaya, 2639.

23. Hearings before the Committee on Commerce, Rivers and Harbors Appropriation Bill, Part 1, March 23–25, 1914, 291; Report of the Board of State Engineers, State of Louisiana, Office Board of Engineers, New Orleans, LA, April 21, 1884, 115; Ibid., April 20, 1896, 7; Ibid., April 20, 1908, 11–12, 94; Riparian Damages on the East Bank of the Mississippi River, Hearings before the United States Congress, March 22, 1912, 55–56, 60.

24. Riparian Damages on the East Bank of the Mississippi River, Hearings before the United States Congress, March 22, 1912, 55–56, 60.

25. *Flood Control and Navigation Maps of the Mississippi River*, Map No. 40; Gauld (Map 1778), "A Plan of the coast of part of West Florida & Louisiana; Wilton (Map 1774), "Part of the River Mississippi from Manchac up to the River Yazous."

26. *Flood Control and Navigation Maps of the Mississippi River*, Map No. 40.

27. Bradford (Map 1838), "Mississippi"; Darby, "Map of the State of Louisiana 1816"; Lucas, (Map 1823), "Mississippi"; Morse and Breese (Map 1842), "Mississippi"; Tanner (Map 1820), "Louisiana and Mississippi"; Norman's Chart of the Lower Mississippi River by A. Persac, published by B.M. Norman, New Orleans, Louisiana 1858; Brooks, "Lloyd's new map of the Mississippi River from Cairo to its mouth," ca. 1862; Cram, "Cram's Railroad & Township Map of Mississippi," 1878.

28. *Flood Control and Navigation Maps of the Mississippi River*, Map No. 40.

29. Ibid.

30. Separation of Red and Atchafalaya Rivers from Mississippi River, Report to the Commission on Rivers and Harbors, House of Representatives, War Department, Washington, DC, March 19, 1914, 5.

31. Some nineteenth-century maps of the area label Bayou Cocodrie as "Bayou Crocodile." See Norman's Chart of the Lower Mississippi River by A. Persac, published by B.M. Norman, New Orleans, Louisiana 1858.

32. Reuss, *Designing the Bayous*, 109; Riparian Damages on the East Bank of the Mississippi River, Hearings before the United States Congress, March 22, 1912, 60.

33. Barry, *Rising Tide*, 426.

34. Noble interview, 130.

35. Törnqvist et al., "A Revised Chronology for Mississippi River Subdeltas," 1694–95.

36. Higby, *Possible Capture of the Mississippi River by the Atchafalaya River*, vi.

37. Ibid., 68–69.

38. Ibid., vi, 68–69.

39. Grady, "Watch on the Mississippi," *Discover*, March 1983, 23–27. Kolb was chief of the Geology Branch, Soils Division, at the Corps' Waterways Experiment Station during 1955–1974. Ben H. Fatheree, "The History of Geotechnical Engineering at the Waterways Experiment Station, 1932–2000," US Army Engineer Research and Development Center, Vicksburg, Mississippi, 2006, Old River Geological Investigations, chap. six, http://gsl.erdc.usace.army.mil/gl-history/Chap6.htm#TORD (accessed October 5, 2015). In 1999, the Corps renamed the Waterways Experiment Station the US Army Engineer Research and Development Center (ERDC) when seven research laboratories were consolidated into one organization. Engineer Research and Development Center: History. US Army Corps of Engineers, http://www.erdc.usace.army.mil/About/History.aspx (accessed August, 14, 2015).

40. Sparks, "Rethinking, Then Rebuilding New Orleans," 1.

41. Ibid., 1, 11.

42. Ibid., 1.

43. Campanella, *Bienville's Dilemma*, 324–25.

44. Mark Fischetti, "Mississippi River Mouth Must Be Abandoned to Save New Orleans from Next Hurricane Katrina," *Scientific American*, August 20, 2015, http://www.scientificamerican.com/article/mississippi-river-mouth-must-be-abandoned-to-save-new-orleans-from-next-hurricane-katrina (accessed September 25, 2015.

45. Sparks, "Rethinking, Then Rebuilding New Orleans," 1, 9–11.

Bibliography

Act No. 581, Regular Session, 2008 Louisiana State Legislature House Bill No. 376 by Representatives Dove and St. Germain, to enact R.S. 38:3097.3 (E), relative to ground water management. http://www.legis.la.gov/legis/ViewDocument. aspx?d=503636 (accessed August 6, 2015.

Act No. 955, Regular Session, 2010 Louisiana State Legislature House Bill No. 1486 (Substitute for House Bill No. 926 by Representative Little) by Representatives Morris et al. relative to the use of surface water. https://www.legis.la.gov/Legis/ ViewDocument.aspx?d=722976 (accessed August 5, 2015).

Acts Passed at the Second Session of the First Legislature of the State of Louisiana, Held and Begun in the City of New Orleans, January 11, 1847 Published by Authority. New Orleans: W. Van Henthuysen, State Printer, 1847.

Acts Passed at the Second Session of the Sixteenth Legislature of the State of Louisiana, 1844.

"After drought, reducing water flow could hurt Mississippi River transport." *New York Times*, November 26, 2012. http://www.nytimes.com/2012/11/27/us/hit-by-drought-mississippi-river-may-face-more-challenges.html?_r=0 (accessed August 28, 2015).

"Agreement on Spillway to Limit Water Flow." *Baton Rouge Morning Advocate*, April 17, 1973.

All-Hazard Authorities of the Federal Emergency Management Agency; the National Flood Insurance Act of 1968, as amended, and the Flood Disaster Protection Act of 1973, as amended; 42 U.S.C. 4001 *et seq.*, Office of the General Counsel, August 1997. https://www.fema.gov/media-library/assets/docu-ments/7277 (accessed March 12, 2016).

Amtrak. http://www.amtrak.com/sunset-limited-train (accessed February 3, 2013).

Anderson, David G., and Kenneth E. Sassaman. "Modeling Paleoindian and Early Archaic Settlement in the Southeast: A Historical Perspective." In *The Paleoindian and Early Archaic Southeast.* Edited by David G. Anderson and Kenneth E. Sassaman, 16–26. Tuscaloosa: University of Alabama Press, 1996.

*The Angola Story: Louisiana State Penitentiary, Angola, Louisiana, Burl Cain,
Warden.* Compiled by Louisiana State Penitentiary Museum Foundation. Rev.
ed., Angola, LA: June 2011.

Annex R, Red River Backwater Area. Mississippi River and Tributaries Comprehensive Review Report, vol. 5, in Mississippi River and Tributaries Project:
Letter from the Secretary of the Army transmitting a letter from the Chief of
Engineers, Department of the Army, dated April 6, 1963, submitting a report,
together with accompanying papers and illustrations, on a review of the Mississippi River and Tributaries Project, in response to a resolution adopted June
12, 1964, by the Committee on Public Works of the US Senate, vol. 5. Washington DC: Government Printing Office, 1964.

Annual Report of the Chief of Engineers on Civil Works Activities, Department
of the Army Corps of Engineers, vol. 1, 1973. Washington, DC: Government
Printing Office, 1973.

Annual Report of the Chief of Engineers on Civil Works Activities, Department
of the Army Corps of Engineers, vol. 1, 1972. Washington, DC: Government
Printing Office, 1972.

Annual Report of the Chief of Engineers on Civil Works Activities, Department
of the Army Corps of Engineers, vol. 2, 1970. Washington, DC: Government
Printing Office, 1970.

Annual Report of the Chief of Engineers, United States Army, on Civil Works
Activities, 1964, in two volumes, vol. 1. Washington, DC: Government Printing
Office, 1965.

Annual Report of the Chief of Engineers, United States Army, to the Secretary of
Defense , for the year 1963, Part 1, vols. 1 and 2. Washington, DC: Government
Printing Office, 1963.

Annual Report of the Chief of Engineers, United States Army, to the Secretary of
Defense, for the year 1961, Part 1, vols. 1 and 2. Washington, DC: Government
Printing Office, 1961.

Annual Report of the Chief of Engineers, United States Army, to the Secretary
of Defense, for the year 1960, vol. 2. Washington, DC: Government Printing
Office, 1960.

Annual Report of the Chief of Engineers, United States Army, to the Secretary of
Defense, for the year 1959, vols. 1 and 2. Washington, DC: Government Printing
Office, 1959.

Annual Report of the Chief of Engineers, United States Army, to the Secretary of
Defense, for the year 1957, Part 1, vols. 1 and 2. Washington, DC: Government
Printing Office, 1957.

Annual Report of the Chief of Engineers, United States Army, to the Secretary of Defense, for the year 1956, Part 1, vols. 1 and 2. Washington, DC: Government Printing Office, 1956.

Annual Report of the Chief of Engineers, United States Army, to the Secretary of Defense, for the year 1953, Part 1, vols. 1 and 2. Washington, DC: Government Printing Office, 1953.

Annual Report of the Chief of Engineers, United States Army, to the Secretary of Defense, for the year 1950, Part 1, vols. 1 and 2. Washington, DC: Government Printing Office, 1950.

Annual Report of the Chief of Engineers, United States Army, to the Secretary of Defense, for the year 1949, Part 1, vols. 1 and 2. Washington, DC: Government Printing Office, 1949.

Annual Report of the Chief of Engineers, United States Army, to the Secretary of War, for the year 1948, Part 1, vols. 1 and 2. Washington, DC: Government Printing Office, 1948.

Annual Report of the Chief of Engineers, United States Army, to the Secretary of War, for the year 1945, Part 1, vols. 1 and 2. Washington, DC: Government Printing Office, 1945.

Annual Report of the Chief of Engineers, United States Army, to the Secretary of War, for the year 1944, Part 1, vols. 1 and 2. Washington, DC: Government Printing Office, 1944.

Annual Report of the Chief of Engineers, United States Army, to the Secretary of War, for the year 1941, Part 1, vols. 1 and 2. Washington, DC: Government Printing Office, 1941.

Annual Report of the Chief of Engineers, United States Army, to the Secretary of War, for the year 1940, Part 1, vols. 1 and 2. Washington, DC: Government Printing Office, 1940.

Annual Report of the Chief of Engineers, United States Army, to the Secretary of War, for the year 1889, in four parts, Part IV. Washington, DC: Government Printing Office, 1889.

Annual Report of the Chief of Engineers, United States Army, to the Secretary of War, for the year 1888, in four parts, Part 4. Washington, DC: Government Printing Office, 1888.

Annual Report of the Chief of Engineers, United States Army, to the Secretary of War, for the year 1885, in four parts, Part 4. Washington, DC: Government Printing Office, 1885.

Annual Report of the Mississippi River Commission for the Fiscal Year ending June 30, 1894; being Appendix XX of the Annual Report of the Chief of Engineers for 1894, Washington, DC: Government Printing Office, 1894.

Annual Report of P. O. Hébert, State Engineer to the [Louisiana] Legislature, January 1847. New Orleans: Magne & Weisse, 1847.

Annual Report of the Secretary of War for the Year 1880. In four volumes; vol. 2, Part 2. Washington: Government Printing Office, 1880.

Annual Report of the State Engineer to the Legislature of the State of Louisiana, February 1846. New Orleans: Jeffersonian Office, 1846.

Aslan, Andres, Whitney J. Autin, and Michael D. Blum. "Causes of River Avulsion: Insights from the Late Holocene Avulsion History of the Mississippi River, U.S.A." *Journal of Sedimentary Research* 75 (2005): 650–64.

"The Atchafalaya." *Times-Picayune* (New Orleans), August 23, 1863.

Atchafalaya Basin Floodway System, Louisiana: Feasibility Study, Technical Appendices A–D. US Army Corps of Engineers, Mississippi River Commission, New Orleans District, January 1982. https://archive.org/stream/Atcha falayaBasinFloodwaySystemLouisianaFeastilityStudyTechAppendices_212/ AtchafalayaBasinFloodwaySystemLouisianaVolume2_djvu.txt (accessed September 16, 2014).

Atkinson, Edward. "Appendix A, Section B" Boston, April 14, 1882. In Mark Twain, *Life on the Mississippi*. By Mark Twain. 1883. Mineola, NY: Dover Publications, 2000, 282–84.

Barnett, James F., Jr. *Mississippi's American Indians*. Heritage of Mississippi Series, vol. 6. Jackson: University Press of Mississippi for the Mississippi Historical Society and the Mississippi Department of Archives and History, 2012.

Barry, John M. *Rising Tide: The Great Mississippi Flood of 1927 and How It Changed America*. New York: Simon and Schuster, 1997.

Bastian, David F. *Grant's Canal: The Union's Attempt to Bypass Vicksburg*. Shippensburg, PA: Burd Street Press, 1995.

"The Bayou Plaquemine." *Times-Picayune* (New Orleans), March 7, 1858.

Begnaud, Allen E. "Acadian Exile." *Journal of the Louisiana Historical Association* 5, no. 1 (Winter 1964): 87–91.

Biedenharn, David S., Colin R. Thorne, and Chester C. Watson. "Recent morphological evolution of the Lower Mississippi River." *Geomorphology* 34 (2000): 227–49.

Bindoff, N. L., J. Willebrand, V. Artale, A. Cazenave, J. Gregory, S. Guley, K. Hanawa, C. Le Quéré, S. Levitus, Y. Nojiri, C. K. Shum, L. D. Talley, and A. Unnikrishnan. "Oceanic Climate Change and Sea Level." In *Climate Change 2007: The Physical Science Basis, Contribution of Working Group I to the Fourth Assessment Report of the Intergovernmental Panel on Climate Change*. Edited by S. Solomon, D. Qin, M. Manning, Z. Chen, M. Marquis, K. B. Averyt, M. Tignor, and H. L. Miller. Cambridge: Cambridge University Press, 2007.

Bonnet Carré's Spillway Overview, New Orleans District, US Army Corps of Engineers, http://www.mvn.usace.army.mil/Missions/MississippiRiverFloodCon trol/BonnetCarréSpillwayOverview.aspx (accessed March 21, 2015).

Bradford, T. G. (Map 1838) "Mississippi" [Boston]. Mississippi Department of Archives and History, MA/92.0078(a).

Bradshaw, Jim. "Great Flood of 1927." *In Know LA: Encyclopedia of Louisiana*. Edited by David Johnson. Louisiana Endowment for the Humanities, 2010–. Article published May 13, 2011. http://www.knowla.org/entry/763/ (accessed October 27, 2014).

Brain, Jeffrey P. *Tunica Archaeology*. Cambridge, MA: Peabody Museum of Archaeology and Ethnology, Harvard University, 1988.

Brain, Jeffrey P., George Roth, and Willem J. De Reuse. "Tunica, Biloxi, and Ofo." In *Handbook of North American Indians*. General editor William C. Sturtevant. Vol. 14, *Southeast*. Edited by Raymond D. Fogelson. Washington, DC: Smithsonian Institution, 2004, 586–97.

Brasseaux, Carl A. *The Founding of New Acadia: The Beginnings of Acadian Life in Louisiana, 1765–1803*. Baton Rouge: Louisiana State University Press, 1987.

Briefing-Mississippi River. Hearing Before the Committee on Public Works, House of Representatives. 93rd Cong., 1st sess. April 11, 1973. Washington, DC: Government Printing Office, 1973.

Brooks, F. W. "Lloyd's new map of the Mississippi River from Cairo to its mouth." H. H. Lloyd & Co., New York, ca. 1862. Library of Congress Geography and Map Division. Catalog Number 99447107. Call Number G4042.M5 1862.B7. http://www.loc.gov/item/99447107 (accessed September 10, 2015).

Broutin, Ignace-François. Carte particulière du cours du fleuve Missisipy ou St. Louis à la Lousiane, depuis la Nouvelle Orléans jusqu'aux Natchez, levée par estime en 1721, 1726, 1731 et dressée au mois d'Aoust 1731. Par Broutin. Bibliothèque nationale de France, Département des cartes et plans. Ge C 5015. http://rla.unc.edu/Natchez (accessed April 9, 2013).

Bureau, Jacques, 1700. Carte du Mississipi. Chicago Historical Society; http://rla.unc.edu/EMAS/regions-ms.html#sec_e (accessed April, 19, 2013).

Burlington Northern Santa Fe Railway. *http://www.bnsf.com/customers/what-can-i-ship/* (accessed February 3, 2013).

Caillot, Marc-Antoine, Erin M. Greenwald, ed., and Teri F. Chalmers, trans. *A Company Man: The Remarkable French-Atlantic Voyage of a Clerk for the Company of the Indies*. New Orleans: The Historic New Orleans Collection, 2013.

Camillo, Charles A. *Divine Providence: The 2011 Flood in the Mississippi River and Tributaries Project*. Vicksburg, MS: Mississippi River Commission, 2012.

Campanella, Richard. *Bienville's Dilemma: A Historical Geography of New Orleans*. Center for Louisiana Studies. Lafayette: University of Louisiana at Lafayette, 2008.

Carleton, Mark T. *Politics and Punishment: The History of the Louisiana State Penal System*. Baton Rouge: Louisiana State University Press, 1971.

Charlevoix, Pierre F. X., Charles E. O'Neill, ed., and John Gilmary Shea, trans. *Charlevoix's Louisiana: Selections from the History and the Journal, Pierre F.X. de Charlevoix*. Baton Rouge: Louisiana State University Press, 1977. Note: Page numbers referenced for Charlevoix follow O'Neill's numbers at the bottom of each page in this edition.

Cheramie, Kristi Dykema. "The Scale of Nature: Modeling the Mississippi River." *Places Journal*, March 2011, https://placesjournal.org/article/the-scale-of-na ture-modeling-the-mississippi-river (accessed January 17, 2015).

Clark, John G. "New Orleans and the River: A Study in Attitudes and Responses." *Journal of the Louisiana Historical Association* 8, no. 2 (Spring 1967): 117–35.

Clauson, Karen. *Measuring Trends in Riverbed Gradation: A Lower Mississippi River Case Study*. Master's thesis, Department of Geography and Environmental Resources, Southern Illinois University in Carbondale, 2009.

Clayton, Lawrence A., Vernon J. Knight, Jr., Edward C. Moore, eds. *The De Soto Chronicles: The Expedition of Hernando De Soto to North America in 1539–1543*. Vols. 1 and 2. Tuscaloosa: University of Alabama Press, 1993.

Coleman, J. F. Consulting Engineer, Atchafalaya Protective Association. Letter to the Mississippi River Commission, April 20, 1911. In Separation of Waters of Red River from Mississippi River. "River Engineering Floods, 1896–1816." Technical Records, A-6 (formerly A-2). General Collection, Mississippi River Commission. Vicksburg, MS.

Colton, Craig E. *An Unnatural Metropolis: Wresting New Orleans from Nature*. Baton Rouge: Louisiana State University Press, 2005.

Comprehensive Report on Reservoirs in Mississippi River Basin: Letter from the Secretary of War transmitting pursuant to Section 10 of the Flood Control Act approved May 15, 1928, a letter from the Chief of Engineers, United States Army, dated July 26, 1935, submitting a report, together with accompanying papers and illustrations, on further flood control of the Lower Mississippi River by control of floodwaters in the drainage basins of the tributaries by establishment of a reservoir system. Washington, DC: Government Printing Office, 1936.

"Conferences to Consider Means of Preventing the Atchafalaya from Capturing the Mississippi." May 19, 1953 (no author). Folder 326, Box 51. Samuel D. Sturgis Jr. Papers. Sturgis Chief of Engineers 1953–1956. Office of History, USACE HQ. Alexandria, VA.

Consultants Conference, Atchafalaya River Study, December 11–13, 1951. Minutes. Research and Technical Records, Library of the Mississippi River Commission. Vicksburg, MS. Provided by IIHR-Hydroscience & Engineering. Iowa City, University of Iowa College of Engineering.

Consultants Conference, Atchafalaya Structure Study, February 9, 1952. Research
and Technical Records, Library of the Mississippi River Commission. Vicksburg, MS. Provided by IIHR-Hydroscience and Engineering. Iowa City, University of Iowa College of Engineering.

Consumer Price Index Inflation Calculator. US Department of Labor, Bureau of
Labor Statistics, http://www.bls.gov/data/inflation_calculator.htm (accessed
July 23, 2015).

Coppola, Damon P. *Introduction to International Disaster Management*. 3rd ed.
Burlington, MA: Butterworth-Heinemann/Elsevier, 2015.

"Corps begins building barrier to block saltwater in Mississippi River." *Fox 8,
WVUE New Orleans*, August 30, 2012, http://www.fox8live.com/story/1930
0284/update-army-corps-begins-building-underwater-barrier-to-block-salt
-water-in-mississippi-river?clienttype=printable (accessed September 5, 2015).

"Corps Plans to Open Spillway at Morganza." *Baton Rouge Morning Advocate*,
April 16, 1973.

"Corps Prepares Morganza Area." *Times-Picayune* (New Orleans), April 11, 1973.

"Corps Raises Backwater Levee." *Concordia Sentinel* (LA), May 16, 1973.

"Corp [sic] Stands Vigil As Mississippi Rises." *Concordia Sentinel* (LA), April 18,
1973.

Costello, Brian J. *A History of Pointe Coupée Parish, Louisiana*. Donaldsonville, LA:
Margaret Media, 2010.

Cram, Geo. F. "Cram's Railroad & Township Map of Mississippi." Published by
Geo. F. Cram, proprietor of the Western Map Depot, 66 Lake Street, Chicago,
IL, 1878. Library of Congress Geography and Map Division, Washington, DC.
Catalog Number 2005625321. Call Number G3981.P3 1878.C7. http://www.loc
.gov/item/2005625321 (accessed September 10, 2015).

CSX Corporation, Railroad Equipment. http://www.csx.com/index.cfm/custom-
ers/equipment/railroad-equipment/#boxcar_specs (accessed March 30, 2015).

Cutrer, Thomas W. "Hébert, Paul Octave." *Handbook of Texas Online*. http://www
.tshaonline.org/handbook/online/articles/fhe09 (accessed July 20, 2014).

Darby, William. "Map of the State of Louisiana 1816." Published by John Melish,
Philadelphia, Pennsylvania. David Rumsey Historical Map Collection. http://
www.davidrumsey.com/luna/servlet/detail/RUMSEY~8~1~2384~220056:
A-Map-of-the-State-of-Louisiana-Wit (accessed September 12, 2015).

Davinroy, Robert D. et al. Hydraulic Sediment Response Modeling, Replication
Accuracy to the River. Technical Paper M53. US Army Corps of Engineers. St.
Louis and Memphis Districts, August 2011. http://mvs-wc.mvs.usace.army.mil/
arec/Documents/Publications/M53_Hydraulic_Sediment_Response_Model
ing_Replication_Accuracy.pdf (accessed July 11, 2015).

Davis, Carolyn Pace. *The Winter of 1863: Grant's Louisiana Canal Expeditions.* The Papers of the Blue and Gray Education Society, No. 4, BGES. Saline, MI: MacNaughton and Gunn, 1997.

Dawdy, Shannon Lee. *Building the Devil's Empire: French Colonial New Orleans.* Chicago: University of Chicago Press, 2008.

Demas, C. R., D. K. Demchek, and Philip B. Curwick. "Sediment Transport in the Lower Mississippi River, 1983–85." In *Proceedings of the Fifth Federal Interagency Sedimentation Conference, 1991.* Edited by Shou-Shan Fan and Yung-Huang Kuo. Vol. 1. Interagency Advisory Committee on Water Data, Subcommittee on Sedimentation, Las Vegas, NV, March 18–21, 1991, 4–79—4–86.

Design of Spillway Tainter Gates. Engineer Manual, Engineering and Design, US Army Corps of Engineers, EM 1110-2-2702, January 2000.

Dewey, Davis R. *Financial History of the United States.* 6th ed. New York: Longmans, Green, 1918.

Dial, D. C., and D. M. Sumner. Geohydrology and Simulated Effects of Pumpage on the New Orleans Aquifer System at New Orleans, Louisiana. Water Resources Technical Report No. 46, Louisiana Department of Transportation and Development in cooperation with the US Geological Survey, 1989.

Dinicola, Karen. *The "100-Year Flood."* US Geological Survey Fact Sheet 229–96; last modified 2/18/2014. http://pubs.usgs.gov/fs/FS-229–96 (accessed October 31, 2014).

"Divorcement of Mississippi and Atchafalaya Rivers. Brief of Atchafalaya Protective Association, submitted to the Mississippi River Commission. 1896–1816." Technical Records, A-2. General Collection, Mississippi River Commission. Vicksburg, MS.

Dockery, David T., III. *Windows into Mississippi's Geologic Past.* Circular 6. Mississippi. Jackson, MS: Department of Environmental Quality, Office of Geology, 1997.

Doolen, Rod. "Batture Lands." *Missouri Conservationist,* September 2, 2005, http://mdc.mo.gov/conmag/2005/09/batture-lands (accessed May 15, 10, 2012).

Draft Programmatic Environmental Assessment, New Orleans Sewerage and Water Board Facilities and Carrollton Water Treatment Plant Hazard Mitigation Proposals. Orleans Parish, Louisiana, HMGP Multiple Projects, FEMA-1603-DR-LA. US Department of Homeland Security, New Orleans, Louisiana Recovery Office, March 2015.

Dredging Activities FY2008-FY2015. New Orleans District, US Army Corps of Engineers, http://www.mvn.usace.army.mil/Missions/Navigation/Dredging Information.aspx (accessed August 31, 2015).

Du Pratz, Antoine Simone Le Page. *The History of Louisiana or of the Western Parts of Virginia and Carolina.* 1774. Reprint, Baton Rouge: Claitor's Publishing Division, 1972.

Durnford, Elias. (Map 1771) "Plan of the River Mississipi [*sic*] from the River Yasous to the River Ibberville in West Florida. Showing lands granted and names of proprietors." With a schedule. [Reproduced in Hulbert (1907–16), ser. 3, plates 97–100]. [*Public Record Office (UK)*, CO 700 / Florida 45.] [Formerly Colonial Office Library (London), Florida 45.]

http://rla.unc.edu/EMAS/regions-ms.html#sec_f (accessed February 14, 2014).

"Economic Impacts of the opening of Bonnet Carré Spillway to Mississippi Oyster Fisheries." Mississippi State University, Coastal Research and Extension Center, http://coastal.msstate.edu/oyster-bonnet-Carré (accessed March 1, 2016).

"Edgar Jadwin, Major General, United States Army." Arlington National Cemetery website. http://www.arlingtoncemetery.net/ejadwin.htm (accessed December 14, 2014).

Einstein, Elizabeth R. *Hans Albert Einstein: Reminiscences of His Life and Our Life Together.* Iowa City: Iowa Institute of Hydraulic Research, the University of Iowa, 1991.

Einstein, Hans Albert. "Appendix II. Bed-Load Measurements in West Goose Creek." In *Bed-Load Transportation in Mountain Creek.* Soil Conservation Service Report SCS-TP-55. Washington, DC: US Department of Agriculture, 1944, 46–50.

Einstein, Hans Albert. Old River Control: Closure of Old River. Water Resources Archives, University of California Riverside and California State University, San Bernadino Libraries, November 1953.

Einstein, Hans Albert. *The Bed-Load Function for Sediment Transportation in Open Channel Flows.* Technical Bulletin No. 1026, September 1950. Washington, DC: USDA Soil Conservation Service, 1950.

Elliott, Charles M., Robert E. Rentschler, and John H. Brooks. "Response of Lower Mississippi River Low-Flow Stages." In *Proceedings of the Fifth Federal Interagency Sedimentation Conference, 1991.* Edited by Shou-Shan Fan and Yung-Huang Kuo. Vol. 1, Interagency Advisory Committee on Water Data, Subcommittee on Sedimentation. Las Vegas, NV, March 18–21, 1991, 4-16–4-19.

Elliott, D. O. *The Improvement of the Lower Mississippi River for Flood Control and Navigation.* 3 vols. Vicksburg, MS: US Waterways Experiment Station, 1932.

Endangered Species Act of 1973. Public Law 95–205, Approved December 28, 1973, 87 Stat. 884 [As amended through Public Law 107–136, January 24, 2002]. http://www.epw.senate.gov/esa73.pdf (accessed July 14, 2015).

Engineer Research and Development Center: History. US Army Corps of Engineers. http://www.erdc.usace.army.mil/About/History.aspx (accessed August 14. 2015).

"Engineers Confident In Backwater Levee." *Concordia Sentinel* (LA), May 2, 1973.

Ettema, Robert, and Cornelia F. Mutel. *Hans Albert Einstein: His Life as a Pioneering Engineer*. Reston, VA: American Society of Civil Engineers, 2014.

Evening Post (NY), August 3, 1831.

Ewens, John. Red River Landing and Head of the Atchafalaya in: Annual Report of the Chief of Engineers, United States Army, to the Secretary of War, for the Year 1884, Part IV, Washington, DC, 2636–41.

Executive Documents of the House of Representatives, Third Session of the Forty-Sixth Congress: 1880–1881. Volume 3, Engineers, no. 1, part 2, vol. 2, part 1. Washington, DC: Government Printing Office, 1881.

Executive Summary. Climate Change Adaption Plan. US Army Corps of Engineers, June 2014, http://www.corpsclimate.us/docs/USACE_Adaptation_Plan _v50_2014_June_lores.pdf (accessed August 24, 2015).

Fairless, Robert T. "The Old River Control Project." In Proceedings: 27th Annual Conference and Symposium. Edited by Dhamo Dhamotharan and Harry C. McWreath. Bethesda, MD: American Water Resources Association, 1991), 261–68.

"Falling Waters." *Times-Picayune* (New Orleans), April 25, 1890.

Fatheree, Ben H. "The History of Geotechnical Engineering at the Waterways Experiment Station, 1932–2000." US Army Engineer Research and Development Center, Vicksburg, Mississippi, 2006; Old River Geological Investigations. Chapter six. http://gsl.erdc.usace.army.mil/gl-history/Chap6.htm#TORD (accessed October 5, 2015).

Federal Emergency Management Agency. Flood Zones. National Flood Insurance Program Policy Index. http://www.fema.gov/floodplain-management/flood -zones (accessed October 31, 2014).

Ferguson, H. B. *History of the Improvement of the Lower Mississippi River for Flood Control and Navigation, 1932–1939*. Vicksburg, MS: US Army Corps of Engineers, Mississippi River Commission 1940.

"Ferriday Chamber of Commerce Fosters Plan to Build Protection Levee." *Concordia Sentinel* (LA), June 25, 1927.

A Few Facts About Melville, Louisiana: The Export Fish Center of Louisiana. Publisher unknown; The Library of Congress, Class F379, Book M5F4, 1882. http:// babel.hathitrust.org/cgi/pt?id=loc.ark:/13960/t18k7kw6n;view=1up;seq=7 (accessed November 1, 2014).

"A Fight With The River." *Times-Picayune* (New Orleans), May 3, 1890.

"Fighting for Protection." *Times-Picayune* (New Orleans), April 23, 1890.

Final Report on Removing Wreck of Battleship "Maine" from Harbor of Habana, Cuba. Washington, DC: US Army Corps of Engineers, 1914.

Fischetti, Mark. "Mississippi River Mouth Must Be Abandoned to Save New Orleans from Next Hurricane Katrina." *Scientific American*, August 20, 2015. http://www.scientificamerican.com/article/mississippi-river-mouth-must-be-abandoned-to-save-new-orleans-from-next-hurricane-katrina (accessed September 25, 2015).

Fisk, Harold N. *Geological Investigation of the Alluvial Valley of the Lower Mississippi River*. War Department, Corps of Engineers, US Army, Vicksburg: Mississippi River Commission Publication No. 52. Vicksburg, MS: Mississippi River Commission, 1944.

Fletcher, B. P., and P. Bhramayana. "Old River Control Auxiliary Structure, Hydraulic Model Investigation." Technical Report HL-88–14, Hydraulics Laboratory, Department of the Army, Waterways Experiment Station, Corps of Engineers. Vicksburg, Mississippi, Final Report, June 1988.

Flood Control Act of 1917. 64th Congress, 2d sess., chaps. 144, 1917. http://www.mvd.usace.army.mil/Portals/52/docs/MRC/Appendix_D_1917_Flood_Control_Act.pdf (accessed January 10, 2015).

Flood Control Act of 1928. 70th Cong., 1st sess., chap. 596, May 15, 1928. http://www.mvd.usace.army.mil/Portals/52/docs/MRC/Appendix_E._1928_Flood_Control_Act.pdf (accessed September 25, 2014).

Flood Control Act of 1938. Public Law No. 761, 75th Cong., 3d sess., chap. 795, H.R. 10618, June 28, 1938. http://www.fws.gov/habitatconservation/Omnibus/FAC1938B.pdf (accessed September 15, 2014).

Flood Control Act of 1954. Public Law No. 780, 83rd Cong., 2d sess., chap. 1264, H.R. 9859, September 3, 1954. http://planning.usace.army.mil/toolbox/library/PL/RHA1954.pdf (accessed 9/16/2014).

Flood Control and Navigation Maps of the Mississippi River, Cairo, Illinois, to the Gulf of Mexico, Including Navigation Charts, Middle Mississippi River Below Hannibal, Missouri. 28th ed. Prepared annually in the office of the president, Mississippi River Commission, Corps of Engineers, US Army. Vicksburg, MS: Mississippi River Commission, 1960.

"Flood Control Ass'n Adopts Resolution." *Concordia Sentinel* (LA), December 12, 1927.

Flood Control in the Lower Mississippi Valley: Report Submitted by the Board of State Engineers to His Excellency, Huey P. Long, Governor of the State of Louisiana, Baton Rouge, November 30, 1929.

Flood Control in the Mississippi Valley. 70th Cong., 1st sess., House of Representatives, Document No. 90, December 8, 1927.

"Flood Control A Profession." *Concordia Sentinel* (LA), April 11, 1973.

Floods and Flood Control on the Mississippi, 1973. The US Army Corps of Engineers, http://babel.hathitrust.org/cgi/pt?id=uiug.30112008443852;view=1up ;seq=1.

"Floodways Explained." *Concordia Sentinel* (LA), March 28, 1973.

"Free Flow Power hydrokinetic turbine deployed in Mississippi River." *Hydro Review* (Tulsa, OK), HydroWorld.com, July 13, 2011. http://www.hydroworld. com/articles/2011/07/free-flow-power-hydrokinetic.html (accessed August 15, 2015).

"Free Flow Power seeks approval for hydropower plants on Allegheny, Monongahela, and Ohio Rivers." *Pittsburgh Post-Gazette*, April 1, 2014. http://power-source.post-gazette.com/powersource/companies/2014/04/01/Free-Flow -Power-seeks-approval-for-10-hydropower-plants-on-Allegheny-Monongahe la-and-Ohio-rivers/stories/201404010019 (accessed August 15, 2015).

Fremling, Calvin R. *Immortal River: The Upper Mississippi in Ancient and Modern Times*. Madison: University of Wisconsin Press, 2005.

Gallay, Alan. *The Indian Slave Trade: The Rise of the English Empire in the American South, 1670–1717*. New Haven, CT: Yale University Press, 2002.

Gauld, George. (Map 1778) "A Plan of the coast of part of West Florida & Louisiana: including the River Yazous / Surveyed by George Gauld M.A. for the Right Honourable the Board of Admiralty." Library of Congress, G4012.C6 1778 .G3 Vault. http://rla.unc.edu/EMAS/regions-ms.html#sec_f (accessed February 14, 2014).

"Gavins Point releases begin scheduled decrease as navigation season closes." Northwestern Division, US Army Corps of Engineers News Release, December 7, 2012. http://www.nwd.usace.army.mil/Media/NewsReleases/tabid/1989/Arti cle/475773/gavins-point-releases-begin-scheduled-decrease-as-navigation-sea son-closes.aspx (accessed September 3, 2015).

Gomez, Gay M. "Describing Louisiana: The Contribution of William Darby." *Journal of the Louisiana Historical Association* 34, no. 1 (Winter 1993): 87–105.

Grady, Denise. "Watch on the Mississippi." *Discover*, March 1983, 23–27.

Graf, Walter H. "Appendix B. Personal Reflections on Hans Albert Einstein's Importance to Hydraulics." In *Hans Albert Einstein: Reminiscences of His Life and Our Life Together*. By Elizabeth R. Einstein (Iowa City: Iowa Institute of Hydraulic Research, University of Iowa, 1991), 101–5.

Graves, E. A. "Hydraulic Requirements." In "Old River Diversion Control—A Symposium." Transaction of the American Society of Civil Engineers, 1958, 1142–59.

Guidance for Incorporating Climate Change Impacts to Inland Hydrology in Civil Works Studies, Designs, and Projects. US Army Corps of Engineers,

Engineering and Construction Bulletin No. 2014-10, May 2014. https://www
.wbdg.org/ccb/ARMYCOE/COEECB/ecb_2014_10.pdf (accessed August 25,
2015).

Gulf South Pipeline Company. Overview.
http://www.gulfsouthpl.com/AboutUsGS.aspx (accessed February 5, 2013).

Perry Gustin. Interviewed by James F. Barnett Jr., January 19, 2015, Morganza, Louisiana. Transcription in the author's research collection, 717 North Pearl Street,
Natchez, Mississippi.

Haffner, Gerald O. "Major Arthur Loftus' Journal of the Proceedings of His Majesty's Twenty-Second Regiment up the River Mississippi in 1764." *Journal of the
Louisiana Historical Association* 20, no. 3 (Summer 1979): 325–34.

Hall, Gwendolyn Midlo. *Africans in Colonial Louisiana: The Development of
Afro-Creole Culture in the Eighteenth Century*. Baton Rouge: Louisiana State
University Press, 1992.

Hardin, John R. "Mississippi-Atchafalaya Diversion Problem." *Military Engineer* 46,
no. 310 (March–April 1954): 87–92.

Harrelson, Danny W. "Geology of Grant's Canal: The Union's Attempt to Bypass
Vicksburg, Mississippi." Paper No. 20–2, Southeastern Section—54th Annual
Meeting, Geological Society of America, Biloxi, MS, March 18, 2005.

Harris, John R. "Alternate Water Sources for the Baton Rouge-New Orleans Indus-
trial Corridor, Addendum A." In "If the Old River Control Structure Fails? The
Physical and Economic Consequences." By Raphael G. Kazmann and David
B. Johnson. *Louisiana Water Resources Research Institute*, Bulletin 12. Baton
Rouge: Louisiana State University, September 1980.

Harrison, Robert W., and Walter M. Kellmorgan. "Socio-Economic History of
Cypress Creek Drainage District and Related Districts of Southeast Arkansas."
Arkansas Historical Quarterly 7, no. 1 (Spring 1948): 20–52.

Harrod, B. M. "The Levees of Louisiana: Their Condition and Requirements."
Times-Picayune (New Orleans), December 6, 1878.

Hearings before the Committee on Commerce, United States Senate, 63rd Cong.,
2d sess. on H.R. 13811. Rivers and Harbors Appropriation Bill, Part 1, March
23–25, 1914. Washington, DC: Government Printing Office, 1914.

Hearings before a Subcommittee of the Committee on Commerce, United States
Senate, 74th Cong., 2d sess. on S. 3531, a bill to amend the act entitled "An Act
for the Control of Floods on the Mississippi River and its Tributaries, and for
other purposes," approved May 15, 1928; January 27–30, 1936. Washington, DC:
Government Printing Office, 1936.

Hearings before the Committee on Commerce, United States Senate, 75th Cong.,
3d sess., on S. 3354, a bill to amend the act entitled "An Act for the Control of

Floods on the Mississippi River and its Tributaries, and for other purposes," approved May 15, 1928; March 28, 29, and 30, 1938. Printed for the use of the Committee on Commerce. Washington, DC: Government Printing Office, 1938.

Hearings before the Committee on Flood Control, House of Representatives, 79th Cong., 1st sess. on H.R. 1902, a bill to amend Section 4 of the act entitled "An Act for the Control of Floods on the Mississippi River and its Tributaries, and for other purposes," approved May 15, 1928; May 25 and October 30, 1945. Washington, DC: Government Printing Office, 1945.

Hearings before the Committee on Flood Control, House of Representatives, 79th Congress, 1st sess. on S. 938, a bill to provide for emergency flood control work made necessary by recent floods, and other purposes, May 14, 1945. Washington, DC: Government Printing Office, 1945.

Hearings before the Committee on Flood Control, House of Representatives, 73d Cong., 1st sess., March 30 and May 12, 1933. Washington, DC: Government Printing Office, 1933.

Hearings before the Committee on Flood Control, House of Representatives, 73rd Cong., 2d sess., or continuation of hearings to provide for lands taken, used, destroyed by reason of setbacks or changes in levee lines on the main channel of the Mississippi River, February 7, 8, and 9, 1934. Washington, DC: Government Printing Office, 1934.

Hebert, Kermit L. "The Flood Control Capabilities of the Atchafalaya Basin Floodway." *Louisiana Water Resources Research Institute, Bulletin GT-1*, April 1967. Baton Rouge: Louisiana State University, 1967.

Higby, John D., Jr. *Possible Capture of the Mississippi River by the Atchafalaya River.* Colorado Water Resources Research Institute, Information Series No. 50, Colorado State University, AE 695V Special Study, August 1983.

Higgs, H. C. *Frequency Curves.* Techniques of Water-Resource Investigations of the Unites States Geological Survey, Chapter A2, Book 4, Hydrologic Analysis and Interpretation, Department of the Interior. Washington, DC: Government Printing Office, 1968.

High Flows and Flood History on the Lower Mississippi River Below Red River Landing, LA (1543–Present). Southern Regional Headquarters, National Oceanic and Atmospheric Administration. http://www.srh.noaa.gov/lix/?n=ms _flood_history (updated October 12, 2011; accessed January 4, 2013).

"History." Louisiana's Old State Capitol. http://louisianaoldstatecapitol.org/Page-Display.asp?p1=805 (accessed July 28, 2014).

"History of the Prison." Angola Museum, Louisiana State Penitentiary Museum Foundation 2013. http://angolamuseum.org/?q=History#history (accessed March 31, 2013).

Hoerling, Martin, Siegfried Schubert, and Kingtse Mo. *An Interpretation of the Origins of the 2012 Central Great Plains Drought.* Assessment Report, NOAA Drought Task Force in partnership with the National Integrated Drought Information System, March 20, 2013. http://www.drought.gov/media/pgfiles/2012-Drought-Interpretation-final.web-041013_V4.0.pdf (accessed August 31, 2015).

Houck, Oliver A. *Down on the Batture.* Jackson: University Press of Mississippi, 2010.

"How the drought on the Mississippi River impacts you." *National Geographic News*, February 1, 2013. http://news.nationalgeographic.com/news/2012/12/121207-nation-mississippi-river-drought-environment-economy (accessed August 28, 2015).

"How would Bonnet Carré Spillway opening impact Pontchartrain fishing?" *Louisiana Sportsman*, January 5, 2016. http://www.louisianasportsman.com/details.php?id=9059 (accessed 3/1/2016).

HSR Modeling Theory. Applied River Engineering Center, St. Louis District, USACE.

http://mvs-wc.mvs.usace.army.mil/arec/HSR_Modeling_Theory.html (accessed July 11, 2015).

Hudson, Charles M. *Knights of Spain, Warriors of the Sun: Hernando de Soto and the South's Ancient Chiefdoms.* Athens: University of Georgia Press, 1997.

Humphreys, A. A., and H. L. Abbott. Report upon the Physics and Hydraulics of the Mississippi River; upon the Protection of the Alluvial Region Against Overflowing; and upon the Deepening of the Mouths: Based upon surveys and investigations made under the Acts of Congress directing the topographical and hydrographical survey of the delta of the Mississippi River, with such investigations as might lead to determine the most practicable plan for securing it from inundation, and the best mode of deepening the channels of the mouths of the river. Submitted by the Bureau of Topographical Engineers, War Department 1861. Washington, DC: Government Printing Office, 1867.

"Hydro christening set." *Concordia Sentinel* (LA), March 29, 1989.

"Hydro-electric plant, Engineers optimistic." *Concordia Sentinel* (LA), February 7, 1978.

"Hydroelectric power viewed." *Concordia Sentinel* (LA), October 11, 1979.

Ingraham, Joseph Holt. *The Southwest by a Yankee.* Vols. 1 and 2. 1835; Readex Microprint Corp., 1966.

"Internal Improvements." *Times-Picayune* (New Orleans), January 19, 1843.

"The Inundation." *Times-Picayune* (New Orleans), April 26, 1874.

"Jazz Festival is Kicked Off." *Times-Picayune* (New Orleans), April 15, 1973.

Jennings, Edward B. "The Life and Death of Old River." *Military Engineer*, July–August 1964, 256–57.

John F. Cubbins, Appt. v. Mississippi River Commission and the Yazoo-Mississippi Delta Levee Board. Supreme Court Reporter, vol. 36, Cases Argued and Determined in the United States Supreme Court, October Term, 1915; December, 1915-July, 1916. St. Paul, MN: West Publishing Company, 1916.

Johnson, David B. "A Change in the Course of the Lower Mississippi River: Description and Analysis of Some Economic Consequences, Addendum B." In "If the Old River Control Structure Fails? The Physical and Economic Consequences." By Raphael G. Kazmann and David B. Johnson.

Johnson, Walter. *Soul By Soul: Life Inside the Antebellum Slave Market*. Cambridge, MA: Harvard University Press, 1999.

Jones, Dennis. *Cultural Resources Survey of Mile 306.3 to 293.4-R on the Mississippi River, Concordia, Pointe Coupee and West Feliciana Parishes, Louisiana*. Cultural Resources Series Report Number: COELMN/PD-91/103; Museum of Geoscience, Louisiana State University, Baton Rouge, Louisiana; Prepared for the US Army Corps of Engineers, New Orleans District; Final Report, August 1993. Baton Rouge, LA: Louisiana State University, 1993.

"Jonesville Hit Hard By Flood." *Concordia Sentinel* (LA), July 9, 1927.

Journal of the Senate of the State of Louisiana, Session of 1848, New Orleans: Office of the *Louisiana Courier*, 1848.

Julian Oliver Davidson: Artist. http://www.battleoflakeerieart.com/jodartist.php (accessed August 16, 2014).

Kansas City Southern Railways. http://www.kcsouthern.com/en-us/Pages/Default.aspx (accessed February 3, 2013).

Kazmann, Raphael G. "Will New Orleans miss the Mississippi?" *Journal of the American Water Works Association*, 73, no. 11 (November 1981): 18, 20.

Kazmann, Raphael G., and David B. Johnson. "If the Old River Control Structure Fails? The Physical and Economic Consequences." *Louisiana Water Resources Research Institute. Bulletin 12*, September 1980. Baton Rouge: Louisiana State University, 1980.

Kelman, Ari. "Boundary Issues: Clarifying New Orleans's Murky Edges." *Journal of American History* 94 (December 2007): 695–703. http://www.journalofamericanhistory.org/projects/katrina/Kelman.html (accessed September 2, 2013).

Klebba, James M. "Water Rights and Water Policy in Louisiana: Laissez Faire Riparianism, Market Based Approaches, or a New Managerialism?" *Louisiana Law Review* 5, no. 6 (July 1993): 1779–1846.

Krinitzsky, Ellis L., and Harold N. Fisk. Geological Investigation of Faulting in the Lower Mississippi Valley. Technical Memorandum No. 3-311, Waterways Experiment Station, Vicksburg, MS, May 1950.

Kunkel, Kenneth E., Laura E. Stevens, Scott E. Stevens, Liqiang Sun, Emily Janssen, Donald Wuebbles, Steven D. Hilberg, Michael S. Timlin, Leslie Stoecker, Nancy E. Westcott, and J. Greg Dobson. Climate of the Midwest, Regional Climate Trends and Scenarios for the US National Climate Assessment, Part 3. US National Oceanic and Atmospheric Administration Technical Report NESDIS 142–3, US Department of Commerce, National Environmental Satellite, Data, and Information Service, Washington DC., June 2013.

Kunkel, Kenneth E., Laura E. Stevens, Scott E. Stevens, Liqiang Sun, Emily Janssen, Donald Wuebbles, Charles E. Konrad II, Christopher M. Fuhrman, Barry D. Klein, Michael C. Kruk, Amanda Billot, Hal Needham, Mark Shafer, and J. Greg Dobson. Climate of the Southeast, Regional Climate Trends and Scenarios for the US National Climate Assessment, Part 2. US National Oceanic and Atmospheric Administration Technical Report NESDIS 142–2, US Department of Commerce, National Environmental Satellite, Data, and Information Service, Washington DC, June 2013.

Kunkel, Kenneth E., Laura E. Stevens, Scott E. Stevens, Liqiang Sun, Emily Janssen, Donald Wuebbles, Devin P. Thomas, Martha D. Shulski, Natalie A. Umphlett, Michael C. Kruk, Kenneth G. Hubbard, Kevin Robbins, Luigi Romolo, Adnan Akyuz, Tapan B. Pathak, Tony Bergantino, and J. Greg Dobson. Climate of the US Great Plains, Regional Climate Trends and Scenarios for the US National Climate Assessment, Part 4. US National Oceanic and Atmospheric Administration Technical Report NESDIS 142–4, US Department of Commerce, National Environmental Satellite, Data, and Information Service, Washington DC, January 2013.

"The Lake Providence Cutoff." *Times-Picayune* (New Orleans), March 29, 1863.

Lamb, J. Parker. *Perfecting the American Steam Locomotive*. Bloomington: Indiana University Press, 2003.

Lamb, Max S., and Loyde T. Ethridge. "Sediment Management on the Mississippi." In *Proceedings of the Fifth Federal Interagency Sedimentation Conference, 1991*, Edited by Shou-Shan Fan and Yung-Huang Kuo. Vol. 1, Interagency Advisory Committee on Water Data, Subcommittee on Sedimentation, Las Vegas, NV, March 18–21, 1991, 1–1—1–8.

Latimer, Rodney A., and Charles W. Schweizer. *The Atchafalaya River Study: A report based upon engineering and geological studies of the enlargement of Old and Atchafalaya rivers including profiles and sections together with factual data which indicate the past rate and extent of progressive changes in Old and Atchafalaya rivers from their head through Grand and Six Mile lakes to the sea, all of which indicate the probable capture of the Mississippi River by the Atchafalaya River*. 3 vols. Vicksburg, MS: Corps of Engineers, Mississippi River Commission, May 1951.

"Letter from Plaquemines." *Times-Picayune* (New Orleans), May 14, 1865.

"The Levees." *Times-Picayune* (New Orleans), December 12, 1874.

Lewis, Pierce F. *New Orleans: The Making of an Urban Landscape.* 2nd ed. Santa Fe, NM: Center for American Places, 2003.

Libby, David J. *Slavery and Frontier Mississippi: 1720–1835.* Jackson: University Press of Mississippi, 2004.

Lindner, C.P. Memorandum Report on Protection of Red River Backwater by a Control Structure in Old River and Operation of Morganza Floodway and Bonnet Carré Spillway. Office of the President, Mississippi River Commission, Vicksburg, Mississippi, August 13, 1945; Research and Technical Records Section, Mississippi River Commission Library, Vicksburg, MS.

"Little Man's Corner, Timber Cut in Flood Area." *Concordia Sentinel* (LA), May 23, 1973.

"Louisiana Legislature." *Times-Picayune* (New Orleans), March 9, 1853.

"Louisiana Legislature." *Times-Picayune* (New Orleans), May 17, 1846.

Lower and Middle Mississippi Valley Engineering Geology Mapping Program. Engineering and Physics Branch, US Army Corps of Engineers, ERDC, Artonish 15 minute quadrangle. http://lmvmapping.erdc.usace.army.mil/Don.htm (accessed May 14, 2015).

Lower Mississippi River Sediment Study. 6 Vols. Catalyst-Old River Hydroelectric Limited Partnership d/b/a Louisiana Hydroelectric Limited Partnership Vidalia, Louisiana in association with US Army Corps of Engineers New Orleans and Vicksburg Districts; US Army Corps of Engineers Waterways Experiment Station Coastal and Hydraulics Laboratory, Vicksburg, Mississippi; Colorado State University Engineering Research Center, Ft. Collins, Colorado; University of Iowa, Iowa Institute for Hydraulic Research, Iowa City, Iowa; Mobile Boundary Hydraulics, Clinton, MS, May 28, 1999; available at the USACE ERDC Library, Vicksburg, MS.

Lucas, Fielding. (Map 1823) "Mississippi" [Baltimore]. Mississippi Department of Archives and History, Archives Library Collection, Jackson, MA/92.0062a.

McCall, Edith. *Conquering the Rivers: Henry Miller Shreve and the Navigation of America's Inland Waterways.* Baton Rouge: Louisiana State University Press, 1984.

McGahey, Samuel O. "A Compendium of Mississippi Dugout Canoes Recorded Since 1974." *Mississippi Archaeology* 21, no. 1 (1986): 58–70.

McGahey, Samuel O. *Mississippi Projectile Point Guide.* Archaeological Report No. 31, Mississippi Department of Archives and History. 2000. Rev. ed. Jackson, MS: Mississippi Department of Archives and History, 2004.

McGahey, Samuel O. "A Prehistoric Dugout Canoe." *Mississippi Archaeology* 9, no. 7 (1974): 24–26.

McPhee, John. "Atchafalaya." In his *The Control of Nature*. New York: Farrar, Straus and Giroux, 1989. 3–92.

McWilliams, Richebourg G., ed. and trans. *Fleur de Lys and Calumet: Being the Pénicaut Narrative of French Adventure in Louisiana*. Tuscaloosa: University of Alabama Press, 1953.

McWilliams, Richebourg G., ed. and trans. *Pierre Le Moyne d'Iberville, Iberville's Gulf Journals*. Tuscaloosa: University of Alabama Press, 1981.

Map of the Cairo & Fulton Railroad Exhibiting the principal tributary lines as projected and its connection with other Railroads west of the Mississippi River, which unite with the Missouri Pacific Railroad and the south projected Pacific Railroad via El Paso to the Pacific Ocean, showing also the Connection by Railroad of the City of New Orleans & St. Louis. Compiled and drawn by I. Wilamowicz, Little Rock, AR, September 1853.

Map: Louisiana Railway Systems, prepared by the Louisiana Department of Transportation and Development, Office of Planning and Programming, Cartography/GIS Unit. *https://www8.dotd.la.gov/estore/products/109-state-maintained-railroad-maps.aspx* (accessed February 3, 2013).

http://usgwarchives.org/maps/louisiana/statemap/la1853.jpg (accessed August 14, 2014).

Map: The Great Overflow, Inundated districts of the Mississippi Valey [*sic*], Compiled and Printed by the *New Orleans Picayune*, 1874. http://usgwarchives.org/maps/louisiana/statemap/1874flood.jpg (accessed August 14, 2014).

Map: Texas Gas Transmission, LLC; Boardwalk Pipeline Partners, LP. http://www.txgt.com/Safety.aspx?id=310; (accessed February 5, 2013).

Maps Showing the Progressive Development of US Railroads, 1830–1950; American Railroads: Their Growth and Development, The Association of American Railroads, 1951; Central Pacific Railroad Photographic History Museum. http://www.cprr.org/Museum/RR_Development.html#1L (accessed July 25, 2014).

Mason, Heather. "Gator Guards Hunted at Angola," WAFB News, West Feliciana Parish, Louisiana, September 15, 2010. *http://www.wafb.com/Global/story.asp?S=13162424&sms_ss=blogger* (accessed June 22, 2013).

Masson, Todd. "Commission changes name of Red River/Three Rivers WMA." NOLA.com/*Times-Picayune*. http://www.nola.com/outdoors/index.ssf/2013/05/commission_changes_name_of_red.html (accessed November 10, 2013).

Mead, Robert H. "Setting: Geology, Hydrology, Sediments, and Engineering of
    the Mississippi River: Water Discharge." US Geological Survey Circular 1133,
    Contaminants in the Mississippi River, Reston, VA, 1995. http://pubs.usgs.gov/
    circ/circ1133/geosetting.html (accessed July 31, 2015).

Meeting to Consider Atchafalaya River Structures, August 25, 1952. Research and
    Technical Records, Library of the Mississippi River Commission, Vicksburg,
    MS; provided by IIHR-Hydroscience & Engineering, University of Iowa Col-
    lege of Engineering.

Memorandum, Gail A. Hathaway, Office of the Chief of Engineers, to Samuel D.
    Sturgis, Chief of Engineers, April 29, 1954. Folder 326, Box 51. Samuel D. Sturgis
    Jr. Papers; Sturgis Chief of Engineers 1953–1956; Office of History, USACE HQ,
    Alexandria, VA.

"Memories are Short—Levee Construction Recalled." *Concordia Sentinel* (LA),
    May 9, 1973.

Minutes of First Meeting of Board of Consulting Engineers on Old River Control,
    April 27, 1954. Folder 326, Box 51, Samuel D. Sturgis Jr. Papers; Sturgis Chief of
    Engineers 1953–1956; Office of History, USACE HQ, Alexandria, VA.

Minutes of Meeting with Consultants on the Study of Means of Preventing
    the Atchafalaya River from Capturing the Mississippi River with Special
    Reference to Structures, September 29–30, 1952. Research and Technical
    Records, Library of the Mississippi River Commission, Vicksburg, MS; pro-
    vided by IIHR-Hydroscience & Engineering, University of Iowa College of
    Engineering.

"The Mississippi." *New York Herald*, July 7, 1830.

"Mississippi and Red River." *Evening Post* (NY), October 17, 1838.

Mississippi River and Tributaries Post-Flood Report, 1973. Department of the Army
    Corps of Engineers, Lower Mississippi Valley Division, Vicksburg, MS.

The Mississippi River and Tributaries Project: Designing the Project Flood. Missis-
    sippi River Commission Information Paper, April 2008.

http://www.mvd.usace.army.mil/Portals/52/docs/Designing%20the%20Project%20
    Flood%20info%20paper.pdf (accessed October 29, 2014).

Mississippi River and Tributaries Project. Mississippi River Mainline Levees,
    Enlargement and Seepage Control. US Army Corps of Engineers, Vicksburg
    District, Vicksburg, MS, Project Report, July 1998.

"Mississippi River Crest Conditions Revised by Officials." *Baton Rouge Morning
    Advocate*, April 2, 1973.

The Mississippi River & Tributaries Project: Birds Point-New Madrid Floodway.
    Mississippi River Commission Information Paper. http://www.semissourian

.com/files/birds-point-new-madrid-info-paper.pdf (accessed November 22, 2014).

Morgan, Arthur E. *Dams and Other Disasters: A Century of the Army Corps of Engineers in Civil Works*. Boston: Porter Sargent, 1971.

Morgan, A. E., S. H. McCrory, and L. L. Hidinger. *A Preliminary Report on the Drainage of the Fifth Louisiana Levee District, Comprising the Parishes of East Carroll, Madison, Tensas, and Concordia*. US Department of Agriculture, Office of Experiment Stations—Circular 104. Washington, DC: Government Printing Office, 1911.

Morgan City 2011: Documenting the Floods for the Morgan City Archives. Tulane University School of Architecture, Service Learning, Fall 2011. http://www.bk.psu.edu/Documents/Academics/Keady-Molanphy_RWC.pdf (accessed March 23, 2015).

"The Morganza Levee Surely Gone." *Times-Picayune* (New Orleans), April 21, 1874.

"Morganza Levees Will Be Bolstered." *Times-Picayune* (New Orleans), April 15, 1973.

"Morganza To Open Tomorrow—Corps." *Times-Picayune* (New Orleans), April 16, 1973.

"Morganza Use Is Reduced to One-Third Its Capacity." *Times-Picayune* (New Orleans), April 17, 1973.

Moore, Jamie, and Dorothy P. Moore. *The Army Corps of Engineers and the Evolution of Federal flood Plain Management Policy*. Program on Environment and Behavior Special Publication No. 20. Boulder, CO: Institute of Behavioral Science, University of Colorado, 1989.

Moore, Norman R. "Structures Required." In "Old River Diversion Control—A Symposium." Transaction of the American Society of Civil Engineers, 1958, 1172–81.

Morris, Christopher. *The Big Muddy: An Environmental History of the Mississippi and Its Peoples from Hernando de Soto to Hurricane Katrina*. New York: Oxford University Press, 2012.

Morris, John W., interviewed by William Settle, 1993, 1995, 1984. Transcript in Research Collections of the Office of History, Headquarters, US Army Corps of Engineers, Alexandria, VA.

Morse, Sidney E., and Samuel Breese. (Map 1842) "Mississippi" [New York]. Mississippi Department of Archives and History, MA/79.0002(a).

Murphy, William L., and Paul E. Albertson. "Engineering Geological Geographical Information System of the Waterways Experiment Station." *Mississippi Geology* 17, no. 2 (June 1996): 36–37.

National Pipeline Mapping system, Public Map Viewer. https://www.npms.phmsa
    .dot.gov/PublicViewer (accessed July 24, 2015).

"Navigation: Ohio and Mississippi Rivers." *Natchez Gazette* (MS), May 29, 1830.

"A new era in energy begins with Vidalia hydro plant." *Concordia Sentinel* (LA),
    October 14, 1985.

Nguyen, Ivan H., Ashley N. Cox, Jasen L. Brown, Robert D. Davinroy, Jason Floyd,
    and Emily Rivera. Mississippi River and Old River Control Complex Sedimen-
    tation Investigation and Hydraulic Sediment Response Model Study. Technical
    Report M53, US Army Corps of Engineers, St. Louis District, Hydrologic and
    Hydraulics Branch, Applied River Engineering Center. Final Report, March
    2011.

Nickles, Charles R., and Thomas J. Pokrefke Jr. "Barge Barrier Study, Technical
    Report HL-84–4, Old River Diversion, Mississippi River, Report 3, Hydraulic
    Model Investigation." Hydraulic Laboratory, US Army Engineer Waterways
    Experiment Station. Vicksburg, MS, 1984.

Nittrouer, Jeffrey A. *Sediment Transport Dynamics in the Lower Mississippi River:
    Non-Uniform Flow and Its Effects on River-Channel Morphology.* PhD disserta-
    tion, University of Texas at Austin, 2010.

"No Opening Planned Yet for Morganza Spillway." *Baton Rouge Morning Advocate*,
    April 3, 1973.

Noble, Charles C., interviewed by Martin Reuss, September 22–23, 1981. Transcript
    in Research Collections of the Office of History, Headquarters. US Army Corps
    of Engineers, Alexandria, VA.

Norman's Chart of the Lower Mississippi River by A. Persac. B. M. Norman, New
    Orleans, Louisiana 1858. Printed and mounted by J. H. Colton, New York.

"No Time To Be Discouraged." *Times-Picayune* (New Orleans), April 23, 1890.

Nussbaum, Patty. *Louisiana Electric Generation—2007 Update.* Technology Assess-
    ment Division, Louisiana Department of Natural Resources. Baton Rouge:
    Louisiana Department of Natural Resources, 2007.

"Object of Note." *Times-Picayune* (New Orleans), August 12, 1866.

"Official Assesses High Water Concern." *Baton Rouge Morning Advocate*, April 14,
    1973.

Old River Control. US Army Corps of Engineers, New Orleans District, January
    2009. N.p.: US Army Corps of Engineers, 1909.

Old River Control Structure Sediment Diversion, Hydraulic Model Investigation,
    Technical Memorandum No. 2–388. Conducted for The President, Mississippi
    River Commission, Vicksburg, MS, by Waterways Experiment Station, June
    1954.

"Old River and the Mud Hole." *Times-Picayune* (New Orleans), September 18, 1877.

"1 million hydro rebate paid." *Concordia Sentinel*, (New Orleans), March 26, 1987.

O'Neill, Karen M. *Rivers by Design: State Power and the Origins of U.S. Flood Control*. Durham, NC: Duke University Press, 2006.

"The Outlets of the Mississippi." *Times-Picayune* (New Orleans), January 21, 1859.

Palfrey, Carl F. "Appendix 3K. Study of Early Maps of the Mississippi River." Annual Report of the Mississippi River Commission for the Fiscal Year Ending June 30, 1893; Being Appendix YY of the Annual report of the Chief of Engineers for 1893. Washington, DC: Government Printing Office, 1893, 3703–8.

Paskoff, Paul F. "Hazard Removal on the Western Rivers as a Problem of Public Policy, 1821–1860." *Journal of the Louisiana Historical Association* 40, no. 3 (Summer 1999): 261–82.

Past MRC [Mississippi River Commission] Members. US Army Corps of Engineers, Mississippi Valley Division.

http://www.mvd.usace.army.mil/About/MississippiRiverCommission(MRC)/PastMRCMembers.aspx (accessed 8/28/2014).

Pearcy, Matthew T. "After the Flood: A History of the 1928 Flood Control Act." *Journal of the Illinois State Historical Society* 95, no. 2 (Summer 2002): 172–201.

Pearcy, Matthew T. "A History of the Randsdell-Humphreys Flood Control Act of 1917." *Journal of the Louisiana Historical Association* 41, no. 2 (2000): 133–59.

Pearson, C. D., and D. G. Hunter. "Moncla Gap and the Red River Diversion." In *Quaternary Geology and Geoarchaeology of the Lower Red River Valley: A Field Trip*. By Whitney J. Autin and Charles E. Pearson. Friends of the Pleistocene, South Central Cell, 11th Annual Field Conference, Alexandria, Louisiana, March 26–28, 1993.

Perrault, S. L., C. E. Pearson, Carey L. Coxe, Sara A. Hahn, Thurston H. G. Hahn III, Dayna Lee, Katherine M. Roberts, and Joanne Ryan. *Archaeological Data Recovery at Angola Plantation, Sites 16WF121 and 16WF122 West Feliciana Parish, Louisiana*. Coastal Environments, Inc. Final Report prepared for New Orleans District US Army Corps of Engineers, 2006. New Orleans: US Army Corps of Engineers, 2006.

Perrault, Stephanie L., Roger T. Saucier, Thurston H. G. Hahn III, Dayna Lee, Joanne Ryan, and Chris Sperling. *Cultural Resources Survey, Testing, and Exploratory Trenching for the Louisiana State Penitentiary Levee Enlargement Project, West Feliciana Parish, Louisiana*. Baton Rouge: Coastal Environments, 2001. http://www.dtic.mil/cgibin/GetTRDoc?AD=ADA387997&Location =U2&doc=GetTRDoc.pdf (accessed June 9, 2014).

Piazza, Bryan P. *The Atchafalaya Basin: History and Ecology of an American Wetland*. Nature Conservancy, 2014. College Station: Texas A & M University Press, 2014.

Prakken, Lawrence B. Groundwater Resources in the New Orleans Area, 2008. Water Resources Technical Report No. 80, US Department of the Interior, US Geological Survey in cooperation with the Louisiana Department of Transportation and Development, Baton Rouge, Louisiana 2009.

"The Raccourci Cut-Off." *New Orleans Commercial Bulletin*, December 24, 1844.

———, November 20, 1844.

———, October 30, 1844.

"Rains Bring Threat of More Floods." *Baton Rouge Sunday Advocate*, April 1, 1973.

Real Estate Handbook. Department of the Army, US Army Corps of Engineers, Washington, DC: US Army Corps of Engineers, November 1985.

"Red River Raft." *Evening Post* (NY), June 9, 1845.

———, June 6, 1839.

———, July 27, 1838.

———, May 23, 1838.

———, June 6, 1837.

Reonas, Matthew. "Delta Planters and the Eudora Floodway: The Politics of Persistence in 1930s Louisiana." *Journal of the Louisiana Historical Association* 50, no. 2 (Spring 2009): 159–87.

"Report made to the Legislature of the State of Louisiana, January 17, 1827, by the Commission of Internal Improvement." *Mississippi Statesman and Natchez Gazette*, May 31, 1827.

Report of a majority of the Committee on the subject of the Raccourci Cut-Off. Louisiana Legislature 1846.

Report of the Board of State Engineers. State of Louisiana, Office Board of Engineers, New Orleans, LA, April 20, 1910, To His Excellency, Jared Y. Sanders, Governor of Louisiana.

Report of the Board of State Engineers. State of Louisiana, Office Board of Engineers, New Orleans, La., April 20, 1908, To His Excellency, Newton C. Blanchard, Governor of Louisiana. Baton Rouge: The Daily State Press, 1908.

Report of the Board of State Engineers. State of Louisiana, Office Board of Engineers, New Orleans, LA, April 20, 1896, To His Excellency, Murphy J. Foster, Governor of Louisiana.

Report of the Board of State Engineers. State of Louisiana, Office Board of Engineers, New Orleans, LA, April 21, 1884, To His Excellency, Samuel D. McEnery, Governor of Louisiana.

"Report of the Chief Engineer to the Secretary of War." *New York Spectator*, December 30, 1831.

Report of the Chief of Engineers, US Army, 1911, in Three Parts, Part 3. Washington, DC: Government Printing Office, 1911.

Report of the Chief of Engineers, US Army, 1926, in Two Parts, Part 1. Washington: DC: Government Printing Office, 1926.

Reuss, Martin. "The Army Corps of Engineers and Flood-Control Politics on the Lower Mississippi." *Journal of the Louisiana Historical Association* 23, no. 2 (Spring 1982): 131–48.

Reuss, Martin. *Designing the Bayous: The Control of Water in the Atchafalaya Basin, 1800–1995.* College Station: Texas A&M University Press, 2004.

Riparian Damages on the East Bank of the Mississippi River. Hearings before the United States Congress, House Committee on the Judiciary. House of Representatives, 62nd Cong., 2d sess., on H.R. 19412, March 22, 1912. Washington, DC: Government Printing Office, 1912.

River Forecast, Lower Ohio/Mississippi River. Lower Mississippi River Forecast Center, National Weather Service, Slidell, LA. http://www.srh.noaa.gov/data/ORN/RVAORN (accessed April 14, 2016).

"River Menace Diminishes." *Times-Picayune* (New Orleans), April 15, 1973.

"The River Problem: Congressman Breckinridge on the Right Treatment of the Mississippi." *Concordia Eagle* (Vidalia, LA), December 13, 1884.

"A River Seceding." *New York Herald*, March 26, 1861.

"The River Situation." *Concordia Sentinel* (LA), April 9, 1927.

River Summary and Forecasts. Vicksburg District Corps of Engineers. http://155.76.244.230/riverstage/bullet.txt (accessed October 6, 2015).

"River Threatens Structure." *Concordia Sentinel* (LA), April 18, 1973.

The *River and Tributaries Project*. US Army Corps of Engineers, New Orleans District. http://www.mvn.usace.army.mil/pao/bro/misstrib.htm (accessed January 2, 2013).

"River Water to Take Toll In Seafood." *Baton Rouge Morning Advocate*, April 9, 1973.

Rivers and Harbors Appropriation Bill, 65th Cong., 2d sess., House of Representatives, Report No. 736, July 2, 1918, Public Law No. 200 [H.R. 10069].

Rollins, Andrew P., Jr., interviewed by Lynn M. Alperin, September 14–15, 1987, Dallas, TX. Transcript in Research Collections of the Office of History, Headquarters, US Army Corps of Engineers, Alexandria, VA.

Room for the River: Summary Report of the 2011 Mississippi River Flood and Successful Operation of the Mississippi River & Tributaries System. N.p.: US Army Corps of Engineers and Mississippi River Commission, 2012.

Ross, Lieutenant. (Map 1775) "Course of the river Mississippi from Balise to Fort Chartres." London: Robert Sayer, 1775. http://www.davidrumsey.com/luna/servlet/detail/RUMSEY~8~1~3664~430011:Course-of-the-River-Mississipi,-fro (accessed June 16, 2014).

Rowland, Dunbar, and Albert G. Sanders, eds. and transls. *Mississippi Provincial Archives, 1704–1743, French Dominion*. Vol. 3. Mississippi Department of Archives and History, Jackson, MS, 1932.

Rowland, Dunbar, and Albert G. Sanders, eds. and transls. *Mississippi Provincial Archives, 1701–1729, French Dominion*. Vol. 2. Mississippi Department of Archives and History, Jackson, MS., 1929.

Rowland, Dunbar and Albert G. Sanders, eds. and transls. *Mississippi Provincial Archives, 1729–1740, French Dominion*, Vol. 1. Mississippi Department of Archives and History, Jackson, MS, 1927.

Rowland, J. C., and W. E. Dietrich. "The Evolution of a Tie Channel." Department of Earth and Planetary Science, University of California-Berkeley. http://eps .berkeley.edu/~bill/papers/rowtet_135.pdf (accessed July 26, 2014). Also in *River, Coastal and Estuarine Morphodynamics: RECM 2005*. Edited by G. Parker and M. Garcia. London: Taylor and Francis, 2006, 725–36.

Sargent, B. Pierre. *Water Use in Louisiana, 2010*. Water Resources Special Report No. 17. Rev. ed. Baton Rouge: Louisiana Department of Transportation and Development, December 2012.

Saucier, Roger T. *Geomorphology and Quaternary Geologic History of the Lower Mississippi Valley*. Vol. 1. US Army Corps of Engineers, Vicksburg, MS, prepared for the President, Mississippi River Commission, 1994. Vicksburg, MS: US Army Corps of Engineers, 1994.

Schneider, David K. "A Matter of Time-Eastern Pipeline Capacity." In *We Are The Practitioners: We are Supply Chain Coaches*, March 26, 2012.

http://wearethepractitioners.com/2012/03/26/a-matter-of-time-eastern-pipeline-capacity (accessed February 5, 2013).

Section 426e (b) Federal Aid in Protection of Shores, 1899 Rivers and Harbors Act.

http://www.gpo.gov/fdsys/pkg/USCODE-2011-title33/pdf/USCODE-2011-title33 -chap9-subchapI.pdf (accessed January 10, 2015).

Separation of Red and Atchafalaya Rivers from Mississippi River. Report to the Commission on Rivers and Harbors, Document No. 841, 63rd Cong., 2d sess., House of Representatives, War Department, Washington, DC, March 19, 1914.

Separation of Waters of Red River from Mississippi River. "River Engineering Floods, 1896–1816," Technical Records, A-6 (formerly A-2), General Collection, Mississippi River Commission. Vicksburg, MS.

Shen, Hseih-Wen. "Appendix C. Hans Albert Einstein's Contributions to Hydraulics." In *Hans Albert Einstein: Reminiscences of His Life and Our Life Together*. By Elizabeth R. Einstein. Iowa City: Iowa Institute of Hydraulic Research, University of Iowa, 1991, 107–9.

Shinkle, Kurt D., and Roy K. Dokka. Rates of Vertical Displacement at Benchmarks in the Lower Mississippi Valley and the Northern Gulf Coast. NOAA

Technical Report NOS/NGS 50. US Department of Commerce. National Oceanic and Atmospheric Administration, National Ocean Service, July 2004.

Simmons, Daryl B., and Fuat Sentürk. *Sediment Transport Technology: Water and
Sediment Dynamics*. Rev. ed. Highlands Ranch, CO: Water Resources Publications, 1992.

Snowdon, J. O., Jr., and Richard R. Priddy. "Geology of Mississippi Loess." *Mississippi Geological, Economic, and Topographic Survey Bulletin* 111 (1968): 13–167.

"Solons Clash on Closing of Structure." *Concordia Sentinel* (LA), April 11, 1973.

South Louisiana Pipelines, Department of Natural Resources, State of Louisiana.
http://dnr.louisiana.gov/assets/docs/oilgas/data/SLA_Pipelines.pdf (accessed
July 24, 2015).

Sparks, Richard E. "Rethinking, Then Rebuilding New Orleans." *Issues in Science
and Technology* 22, no. 2 (Winter 2006): 1–12. http://issues.org/22-2/sparks
(accessed September 24, 2015).

Special Report of the Mississippi River Commission on Revision of the Plans
for Improvement of Navigation and Flood Control of the Mississippi River.
November 28, 1927, St. Louis, MO. 70th Cong., 1st sess., Senate Committee.
Washington, DC: Government Printing Office, 1928.

"Spillway Opened Above New Orleans." *Baton Rouge Morning Advocate*, April 9,
1973.

"Startups explore alternative hydro power on the Mississippi." *Midwest Energy
News* (St. Paul, MN), June 21, 2011.

http://midwestenergynews.com/2011/06/21/startups-explore-alternative-hydro
-power-on-the-mississippi (accessed August 13, 2015).

Stephenson, Wendell Holmes. *Isaac Franklin: Slave Trader and Planter of the Old
South, With Plantation Records*. 1938; Gloucester, MA: Peter Smith, 1968. Stickney, Amos. Red and Atchafalaya Rivers, Progress of Surveys and Examinations.
In Annual Report of the Chief of Engineers, United States Army, to the Secretary of War, for the Year 1884, Part 4, Washington, DC, 2419–21.

Stoddard, Amos. *Sketches of Louisiana, Historical and Descriptive*. 1812; Reprint,
Carlisle, MA: Applewood Books, n.d.

Streever, Bill. *Saving Louisiana: The Battle for Coastal Wetlands*. Jackson: University
Press of Mississippi, 2001.

Swanson, Mark T. Interpretive Program, Plaquemine Locks State Commemorative Area. Project No. 06-06-00-78-12. New World Research, Inc. Report of
Investigations No. 82-30. State of Louisiana, Office of State Parks, November
28, 1983.

Swanton, John R. *Indian Tribes of the Lower Mississippi Valley and Adjacent Coast
of the Gulf of Mexico*. Smithsonian Institution Bureau of American Ethnology
Bulletin 43. 1911; Reprint, Mineola, NY: Dover Publications, 1998.

Tanner, Henry S. (Map 1820) "Louisiana and Mississippi" [Philadelphia]. Mississippi Department of Archives and History, MA/87.0006(c).

Technology: 1828–1840, Cleaning up the Mississippi. Illinois State Museum. http://
  www.museum.state.il.us/RiverWeb/landings/Ambot/TECH/TECH12.htm
  (accessed April 26, 2014).

Tensions rise over salt water intrusion into BR aquifers." *Baton Rouge Advocate*,
  May 5, 2014. http://theadvocate.com/home/9057272–125/frustration-mount
  ing-over-saltwater-intrusion (accessed July 31, 2015).

Thomas, E. J., letter to Charles L. Potter, USACE and Secretary, Mississippi River
  Commission, May 15, 1912. In "Separation of Waters of Red River from Mississippi River. River Engineering Floods, 1896–1816." Technical Records, A- (formerly A-2), General Collection, Mississippi River Commission, Vicksburg, MS.

Thomas, E. J. Report on Survey of Vicinity of Mouth of Red River, La., 1910–1911,
  with estimates of Cost of Closure Works. In Separation of Waters of Red River
  from Mississippi River. "River Engineering Floods, 1896–1816," Technical
  Records, A-6 (formerly A-2), General Collection, Mississippi River Commission, Vicksburg, MS.

Thompson, Jeff. "Letter from Gen. Jeff Thompson." *Times-Picayune* (New Orleans),
  May 20, 1876.

Thorne, Colin, Oliver Harmer, Chester Watson, Nick Clifford, David Biedenharn,
  and Richard Measures. *Current and Historical Sediment Loads in the Lower
  Mississippi River*. Final Report to United States Army, European Research
  Office of the US Army, Contract Number 1106-EN-01, School of Geography,
  University of Nottingham, University Park, Nottingham NG7 2RD, July 2008.
  Nottingham, UK: US Army Corps of Engineers, 2008.

Törnqvist, Torbjörn E., and John S. Bridge. "Spatial variation of overbank aggradation rate and its influence on avulsion frequency." *Sedimentology* 49 (2002):
  891–905.

Törnqvist, Torbjörn E., Tristram R. Kidder, Whitney J. Autin, Klaas van der Borg,
  Arie F. M. de Jong, Cornelis J. W. Klerks, Els M. A. Snijders, Joep E. A. Storms,
  Remke L. van Dam, and Michael C. Wiemann. "A Revised Chronology for
  Mississippi River Subdeltas." *Science*, new ser., 273, no. 5282 (September 20,
  1996): 1693–96.

Turnbull, Willard J., and Woodland G. Shockley. "Foundation Design." In "Old
  River Diversion Control—A Symposium." *Transactions of the American Society
  of Civil Engineers*, 1958, 1160–71.

"Trip on the Atchafalaya River—Drawn by J. O. Davidson." *Harper's Weekly*, April
  14, 1883, 237.

Twain, Mark. *Life on the Mississippi*. 1883. Mineola, NY: Dover Publications, 2000.

2011 Flood Fight. US Army Corps of Engineers, New Orleans District. http://www.mvn.usace.army.mil/bCarré/floodfight.asp (accessed January 2, 2013).

"2012 drought impacting Mississippi River barges." *AccuWeather.com*, July 25, 2012. http://www.accuweather.com/en/weather-news/drought-impacting-mississippi-barges/67963 (accessed August 28, 2015).

"Underwater dam in Mississippi River still protecting New Orleans, Jefferson water supplies from saltwater as drought continues." *Times-Picayune* (New Orleans), NOLA.com, December 17, 2012. http://www.nola.com/environment/index.ssf/2012/12/underwater_dam_in_mississippi.html (accessed September 7, 2015).

Union Pacific in Louisiana, 2011 Fast Facts, Revised 3/12. http://www.up.com/cs/groups/public/documents/up_pdf_nativedocs/pdf_louisiana_usguide.pdf (accessed February 3, 2013).

US Energy Information Administration. Natural Gas. US Department of Energy, Washington, DC. http://www.eia.gov/pub/oil_gas/natural_gas/analysis_publications/ngpipeline/southwest.html (accessed February 5, 2013).

US EPA Office of Wetlands, Oceans, and Watersheds. http://water.epa.gov/type/watersheds/named/msbasin/marb.cfm (accessed February 7, 2015).

Usner, Daniel H., Jr. "From African Captivity to American Slavery: The Introduction of Black Laborers to Colonial Louisiana." In *The Louisiana Purchase Bicentennial Series in Louisiana History*. Vol. 1: *The French Experience in Louisiana*. Edited by Glenn R. Conrad. Lafayette: Center for Louisiana Studies, University of Southwestern Louisiana, 1995. 183–200.

Van Arsdale, Roy B. *Adventures Through Deep Time: The Central Mississippi River Valley and Its Earthquakes*. Geological Society of America, Special Papers 455, 2009. Boulder, CO: GSA, 2009.

van Beek, Johannes L. Hydraulics of the Atchafalaya Basin Main Channel System: Consideration from a Multiuse Management Standpoint. Contract No. 68–03–2665. Environmental Monitoring and Support Laboratory, Office of Research and Development, US Environmental Protection Agency, Las Vegas, NV, May 1979.

van Beek, Johannes L., Ava L. Harmon, Charles L. Wax, and Karen M. Wicker. Operation of the Old River Control Project, Atchafalaya Basin: An Evaluation from a Multiuse Management Standpoint. Contract No. 68–03–2665. Environmental Monitoring and Support Laboratory, Office of Research and Development, US Environmental Protection Agency, Las Vegas, Nevada, November 1979.

van Beek, Johannes L., Karen Wicker, Benjamin Small. A comparison of Three Floodway Regimes, Atchafalaya Basin, Louisiana. Contract No. 68–01–2299. Environmental Monitoring and Support Laboratory, Office of Research and Development, US Environmental Protection Agency, Las Vegas, Nevada, December 1978.

Vertical Datums. National Geodetic Survey, National Oceanic and Atmospheric Administration. http://www.ngs.noaa.gov/datums/vertical (accessed 7/7/2015).

"Vidalia may still be facing a multi-million deficit." *Natchez Democrat* (MS), September 23, 2012.

"Vidalia, Police Jury join forces to seek industry." *Concordia Sentinel* (LA), November 29, 1977.

"Vidalia power plant nearer to reality." *Concordia Sentinel* (LA), February 23, 1978.

Wardin, Albert W. Jr. *Belmont Mansion: The Home of Joseph and Adelicia Acklen.* Nashville, TN: Belmont Mansion Association, 2005.

"The Watergate Story." *Washington Post.* http://www.washingtonpost.com/wp-srv/politics/special/watergate/timeline.html (accessed May 22, 2015).

Wernet, Mary Linn. "The United States Senator Overton Collection and the History It Holds Relating to the Control of Floods in the Alluvial Valley of the Mississippi, 1936–1948." *Journal of the Louisiana Historical Association* 46, no. 4 (Autumn 2005): 449–64.

"What the Bonnet Carré Spillway opening means for Lake Pontchartrain." *NOLA.com/ Times-Picayune* (New Orleans), January 8, 2016. http://www.nola.com/environment/index.ssf/2016/01/bonnet_Carré_spillway_opening.html (accessed March 1, 2016).

White, Gilbert F. *Human Adjustment to Floods: A Geographical Approach to the Flood Problem in the United States.* Research Paper No. 29. Department of Geography Research Papers. Chicago: University of Chicago, 1945.

Whittington, Mitchel. *No Hope! The Story of the Great Red River Raft.* St. Francisville, LA: 23 House Publishing, 2009.

Williams, Stephen, and Jeffrey P. Brain. *Excavations at the Lake George Site, Yazoo County, Mississippi, 1958–1960.* Papers of the Peabody Museum of Archaeology and Ethnology. Vol. 74. Cambridge: Papers of the Peabody Museum of Archaeology and Ethnology, 1983.

Willis, Homer, interviewed by Bruce Kalk, March 15, 1991, Bethesda, Maryland. Transcript in Research Collections of the Office of History, Headquarters, US Army Corps of Engineers, Alexandria, VA.

Wilson, Samuel Jr. "Colonial Fortifications and Military Architecture in the Mississippi Valley." In *The Louisiana Purchase Bicentennial Series in Louisiana History.* Vol. I: *The French Experience in Louisiana.* Edited by Glenn R. Conrad.

Lafayette: Center for Louisiana Studies, University of Southwestern Louisiana, 1995, 378–94.

Wilton, William. (Map 1774) "Part of the River Mississippi from Manchac up to the River Yazous." [2 sheets, 27.25" x 67.25".] [Mississippi River Commission (Vicksburg, MS), 91.]

http://rla.unc.edu/EMAS/regions-ms.html#sec_f (accessed February 14, 2014).

Woo, Hyoseop S., Pierre Y. Julien, and Everette V. Richardson. "Washload and Fine Sediment Load." *Journal of Hydraulic Engineering* 112, no. 6 (June 1986), © American Society of Civil Engineers, http://www.engr.colostate.edu/~pierre/ce_old/Projects/Paperspdf/Woo-Julien-Richardson-86.pdf. 541–45.

Yodis, Elaine G., Craig E. Colten, and David C. Johnson. *Geography of Louisiana.* Boston: McGraw-Hill, 2003.

Index